GEOLOGY IN THE
NINETEENTH CENTURY

CORNELL HISTORY OF SCIENCE SERIES

Editor: L. Pearce Williams

Norma Emerton, *The Scientific Reinterpretation of Form*

GEOLOGY IN THE NINETEENTH CENTURY

CHANGING VIEWS OF A CHANGING WORLD

By MOTT T. GREENE

CORNELL UNIVERSITY PRESS
Ithaca and London

To my parents,
DR. LYNNE T. GREENE
and IRENE B. GREENE

Foreword

Iт is fit and proper that a series in the history of science should be published by Cornell University Press. Cornell, together with Harvard and the University of Wisconsin, was one of the original centers from which the discipline of history of science grew after World War II. Under the guidance of Henry Guerlac, there developed what can be called the "Cornell school" of the history of science, which has been marked by a certain daring in its selection of topics and an elegance in the presentation of detailed historical research. The Cornell History of Science Series will attempt to preserve and forward these characteristics.

Although the discipline of the history of science is now a mature one, it is also relatively undeveloped. There are whole areas that remain untouched, scientific lives that remain unexamined, and relations between science and society that are still unsuspected. The Cornell History of Science Series is intended as a means for the publication of works that deal with concerns central to the development, expansion, and use of scientific ideas, theories, and practices. Works published in the series will be of the highest scholarly quality, and they will focus upon matters of fundamental concern, not just for the scholar, but for all who are interested in the life of science. They will be written in a clear and straightforward prose. When and where technical jargon is required, there will be glossaries to acquaint the nonspecialist with the language of the subject, and that language will be used always to enlighten, not to obscure, the ideas behind it.

Mott Greene's work exemplifies these qualities. The history of geology in the nineteenth century has, for too long, been seen as the conflict between Werner and Hutton, neptunists and vulcanists, with Charles Lyell emerging as the great conciliator who set everything aright. Professor Greene shows that this view, largely the work of Lyell himself, is profoundly wrong and that the major thread of geological thought ran in a different direction. Geologists were far more concerned with the actual ways in which the face of the earth was formed, particularly with how mountains arose, than could be gathered from Lyell's account of things. In Professor Greene's perspective, it is the continental geologists who occupy the center of the stage. His historical spotlight shines upon them and reveals, for the first time, their basic contributions to our knowledge of the earth.

L. PEARCE WILLIAMS

Ithaca, New York

Contents

[9]

Contents

Illustrations

Preface

THIS book began as an examination of late-nineteenth-century geology, an outgrowth of my interest in the character of scientific inquiry during a period in which, as far as I could gather from historical accounts, no major theoretical shifts occurred. I was interested in what has been referred to, in a well-known theory of scientific change, as "normal science" and, in another such theory, as the "elaboration of the research programme": the day-to-day application of established principles of interpretation as a means to clearly defined goals. In geology this meant the application of the theories of uniform geological change and of biological evolution to the elucidation of stratigraphy and life history. The goal of this enterprise was a coherent historical account of the transformation of the face of the earth and of the evolutionary modification of life forms conditioned by the slow but perpetual modification of the environment.

I had hoped to study the process of inductive generalization used in elaborating such a theory: the sifting and coordination of research reports as more and more of the earth came under the geologist's eye, hammer, and microscope; the subtle interplay of controversy and rival interpretation; the niceties of taxonomy and classification; the methodical and cooperative work of a well-organized, well-funded branch of scientific investigation.

Instead, I found a science deep in controversy, divided on issues of fact, interpretation, and method and even on the aim and scope proper to geology as a science. I found that continental European geologists proudly traced their lineage back to figures dismissed in English-language histories as delirious fantasts; conversely, the heroes of British geology were either ignored on the continent because of the involvement of their geology with theology or pitied for their inability to comprehend the fundamental importance of mountain building and episodic disturbance. I discovered that North American geologists generally found the geology of the Old World as odious as its politics and fashioned American theories to solve American problems.

But I found something more. I learned that in the late nineteenth century, German, Austro-Hungarian, French, and Swiss geologists, joined by a minority of British and Americans at work on the same problems, had fashioned a powerful and comprehensive theory of earth history, which for a short time had come close to unifying geologists into a single community. Essentially a catastrophic theory, it divided earth history into periods of violent disturbance and quiet readjustment. It specified the coherent lateral movement of sections of the earth's crust as wide as the Alps for dozens and eventually hundreds of kilometers and developed a substantial body of evidence indicating former continuity across the Atlantic abyss. This continuity was demonstrated on structural, stratigraphic, petrologic, and paleontological grounds and dynamically interpreted in terms of the earth's nebular origin and subsequent cooling—a cooling that progressively destroyed continental crust and pointed inexorably to a future static earth covered by a universal ocean.

Proposed in the middle of the nineteenth century, the theory became prominent by the late 1880s and had a worldwide following by 1900. Its decline was even more rapid than its growth, however, and by 1910 it was everywhere under attack. Its collapse was followed by a period of theoretical pluralism, confusion, and even despair.

The standard history of geology offered in the English-speaking world is markedly incomplete. Based on principles of continental permanence originally developed in the first half of the nineteenth century and successively expanded thereafter into the first half of the twentieth, and faced with an increasingly disturbing and extensive body of anomalous facts only recently reorganized and integrated by plate tectonics, the standard account omits an important chapter in the history of geology and gives a misleading picture of the nature and frequency of major theoretical shifts: change in geological theory is not slow, cautious, and uniform, but rapid, radical, and sometimes catastrophic.

What follows is an examination of major controversies in geology since the late eighteenth century. The usual overemphasis on the importance of certain British geologists has been corrected in these pages by the suggestion that several influential geologists from continental Europe, whose careers seem aberrant in the context of the steady elaboration of essentially British insights into a universally accepted theory of continental permanence, may now find a comprehensible place in the evolution of geological theory.

The thesis I argue is not intended as a polemic, and I offer the outlines of a history, not a theory, of geology. My aim at every turn has been to provide an appreciation of the manifold difficulties involved in obtaining agreement on a general theory and to show that while major controversies may have clarified issues, they rarely produced a determinate outcome in which one side went down to defeat and extinction and the other rose to the status of ruling theory. No credence can be given to the idea that geology has advanced and changed in a series of clear-cut victories of one theory over another, followed by a period of unanimity and orthodox interpretation. In no period from the eighteenth century to the present has such a state of affairs existed across all national boundaries.

I have also argued (implicitly) that a clear and comprehensive thesis, even if it is generally and contemporaneously viewed as false, can often be more valuable and influential than one that is vague, narrow, and obtuse but finds universal contemporary assent.

I commend this picture of the history of geology to several audiences. I hope that the general reader will share with me the excitement of geology and geological theory as an intellectual quest. I hope that historians of science will find it useful. I hope particularly that geologists will find evidence of the great benefits and equally great dangers involved in unswerving allegiance to a general theory and will find as well an appreciation of the unique character of geology as a scientific enterprise.

I am grateful to Thomas L. Hankins, Carol Thomas, and D. James Baker of the University of Washington for their criticism and advice on an early version of this work. Later I received valuable comments from Henry Frankel of the University of Missouri

(Kansas City), Rachel Laudan of the University of Pittsburgh, Stephen Brush of the University of Maryland, and Harold Burstyn, historian of the U.S. Geological Survey. I owe particular thanks to L. Pearce Williams of Cornell University, editor of this series, for his encouragement, guidance, and help.

I feel a special debt to three scientist-historians: Gordon Davies, Helmut Hölder, and Ursula Marvin. Their excellent work in the history of geology—Davies on the history of geomorphology, Hölder on recurrent issues in tectonics, and Marvin on the origin of continental drift theory—had a strong influence on my own views.

Julian Barksdale, of the Department of Geology, University of Washington, provided much information on the history of orogenic theory and the career of Bailey Willis. Peter Misch, of the same department, allowed me to attend his graduate course in goetectonics in the fall of 1979 and provided a wealth of geological and historical insight. As a teacher, geologist, and mountaineer, he has carried the grand traditions of European tectonics to the Pacific Northwest. My acquaintance with him has made this a better book.

The governing board of the Bonnet Institute, particularly Lorin Anderson of Portland, Oregon, and Claude Singer of New York City, knows my intellectual debt to the institute and its members, but I am happy to have the opportunity to express my thanks here, as well.

I wish to thank Oxford University Press for permission to use figures from E. B. Bailey's *Tectonic Essays, Mainly Alpine* (1935).

Last, and far from least, I am grateful to Jo Leffingwell, who badgered me unmercifully through the process of revision and without whose prodding and encouragement I could not have finished this work.

NOTE TO THE 1984 PRINTING

My deep thanks to those reviewers and correspondents who have expressed interest in this book and offered generous criticisms to the author.

I stand by all my interpretations and emphases, with two notable exceptions. If I were to rewrite the book, I would give more attention to Cuvier and the impact of his work, which I seriously underestimated. I would also give less credit to Maurice Lugeon, and a great deal more to Hans Schardt for the unravelling of the mystery of the structure of the Pre-Alps.

MOTT T. GREENE

Bainbridge Island, Washington

GEOLOGY IN THE
NINETEENTH CENTURY

CHAPTER ONE

Hutton and Werner: First Principles

Two important histories of geology appeared at the beginning of the twentieth century. The first was Karl von Zittel's *History of Geology and Paleontology to the End of the Nineteenth Century*, translated (in abridged form) from German into English (Zittel 1901). It was much as the title suggests, a scholarly and compendious chronicle of every contribution to either science that seemed to lead anywhere at all; it is still the most comprehensive attempt at a history of geology. The other was Sir Archibald Geikie's *Founders of Geology*, which, also true to its title, is an epitome of biographical sketches of those geologists who, in Geikie's estimation, had contributed most to the establishment of geology in its modern form (Geikie 1905). In spite of the difference in their historical approaches, both writers stressed the revolutionary importance of the work of James Hutton (1726–1797) of Edinburgh in shaping the development of geological theory in the nineteenth century. Both Zittel and Geikie attributed to Hutton a number of key generalizations that focused the attention of geologists on—and in some instances solved—the most significant problems to be considered. Among them were Hutton's postulate of limitless geological time and his insistence that "the present is the key to the past": that is, that the earth changes and is changed today by the same agencies and processes that have always acted, and that a close study of the present alterations of the earth can explain past geological history. Hutton recognized more fully than any of his contemporaries the central role of the earth's internal heat in creating and transforming rocks and minerals. He outlined the life cycle of continents and oceans as a ceaseless rhythm of erosion and deposition, in which continents were slowly worn down by running water and were carried to the sea as sediments, later to be elevated as new continents.

This picture of Hutton as the true founder of modern geology remained intact in the next major history of geology to appear in English, Frank D. Adams's *The Birth and Development of the Geological Sciences* (1938). Adams recognized that Hutton had turned the attention of geologists from deluges and other catastrophes toward a reasoned examination of the succession of small events that, through great stretches of time, had wrought all the changes on the face of the earth. Hutton's primacy was reinforced in similar terms more recently by the publication of E. B. Bailey's *James Hutton: The Founder of Modern Geology* (1967).

In all of these works, Hutton's geological theories are presented as the result of long years of observation. Isolated at first from other geologists, he discovered independently much of the geological knowledge laboriously deduced by a host of investigators in previous centuries. For example, he observed that the presence of fossil shells in continental rocks must indicate their marine origin and that some rocks were not massive and crystalline but composed of sand, mud, and other eroded elements of older rocks. He rediscovered the principle of superposition—the lower rocks must be the older—and developed a means of determining whether a given sequence of sedimentary beds was in its original position or had been overturned and therefore reversed. He saw that sedimentary rocks had been invaded, crosscut, and otherwise disrupted by the injection of molten rock after their original deposition, and he perceived that rocks could be so altered by heat that they lost all or nearly all of their original character.

Hutton interpreted the sinuous flexures of many sedimentary beds as evidence of the action of heat and pressure while the sediments were deeply buried. He observed that horizontal beds of sediment were sometimes contiguous with steeply tilted beds and inferred that the geological record was not complete and that periods of deformation and erosion had occurred between successive eras of deposition. All of these matters were systematically presented by Hutton in his *Theory of the Earth with Proofs and Illustrations,* which may be regarded as the just cause of his fame (1795).

Over and above the content and character of his geological work, certain elements of Hutton's life accentuate his stature. He was part of a brilliant circle that flourished in Scotland at the end of the eighteenth century and was a friend and intimate of Joseph Black, James Hall, Adam Smith, Adam Ferguson, James Clerk, and other men who contributed to the physical sciences, economics, history, sociology, military strategy, and philosophy.

Moreover, Hutton tried to integrate his work in geology with natural philosophy in general. Several historians have recently suggested that his treatises in natural philosophy and on the theory of heat and matter should be viewed as part of an important tradition within the "Scottish Philosophy" that led to the work of William Thomson and James Maxwell.[1] In this work he joined many later geologists (including Geikie and Zittel) who believed that geology would be fully integrated with natural science and placed on the firm footing of physics and chemistry.

Hutton's fame is a well-defended fortress. Much of his work was original; what was not original was sound or at least informed with the prescient recognition that geological science should properly dissolve itself into the basic physical sciences as soon as possible. Independent, inventive, philosophically minded, he consistently directed his work toward issues in basic science, combining the best of contemporary observation into a useful synthesis. Hutton, who was linked closely to the general advance of the intellectual life of his time, has been hurled by this torrent of excellence toward an inevitable apotheosis.

Some of the legend surrounding Hutton is true, but his fame is like a geode, and if we split the solid mass apart we see a nodular growth, successive concretions around a hollow center, varicolored, each layer representing the needs and aims of the geologists who contributed to the legend. For with the postulation of a revolutionary figure there comes a historiographic pressure to show the clear superiority of the revolutionary view over everything that preceded it. In Hutton's case, this need has been achieved by exaggerating his originality and suppressing, in successive accounts, those elements in his system that seem backward-looking and wrong while at the same time accentuating those very elements in the work of his opponents. The result is a historical account that systematically

---

1. For various aspects of this question see Gerstner 1968; Heimann and McGuire 1971:281ff.; Olson, 1969 and 1975. Whether anyone actually read Hutton's treatises is another matter. James McCosh (1875:262) described Hutton's *Investigation of the Principles of Knowledge* as follows: "The work is full of awkwardly constructed sentences and of repetitions, and it is a weariness in the extreme to read it. Yet we are made to feel at times that these thoughts must be profound, if only we could understand them."

distorts our view of the state of the earth sciences at the beginning of the nineteenth century.

In 1785, Hutton read a paper before the Royal Society of Edinburgh, "On the Theory of the Earth; or, An Investigation of the Laws Observable in the Composition, Dissolution, and Restoration of the Globe." The paper was published three years later in the first volume of the society's proceedings (Hutton 1788). In 1795, Hutton produced a much larger work, his *Theory of the Earth,* in which he made some attempt to answer his critics and expanded the consideration of the evidence for his theory, largely by the inclusion of long extracts (still in French) from Horace Bénédict de Saussure's *Voyages dans les Alpes* (1779–1796). Hutton died in 1797, and the defense of the theory passed to his friend, John Playfair, who published in 1802 the *Illustrations of the Huttonian Theory of the Earth,* which became the revised standard version of the Huttonian theory and the vehicle of its subsequent popularity and notoriety (Playfair 1802).

Playfair tried in this work to present Hutton's theory in a coherent and readable form and to defend it from theological criticism by divorcing Hutton's geological ideas from natural theology. He also put forward a historical and philosophical interpretation of Hutton's work as a qualitative departure from all previous theories, representing it as a "Newtonian revolution" in geological thought.

Hutton's own work was turgid and diffuse, his style labyrinthine and indecisive, full of compound, complex constructions in which digression and reservation won out consistently over the intended statement. It was also a paean to the beneficent intentions of the Creator in devising such a harmonious system as our earth. Hutton's theory of the earth was a classic of deistic natural theology and natural philosophy in which geological phenomena were evidence for the conclusion that no system so balanced and purposeful could ever have emerged without intelligent design.

Those familiar with natural philosophy in the seventeenth and eighteenth centuries are aware that the dividing line between it and natural theology is often narrow and as often invisible. Isaac Newton's work is a supreme example of the organic connection between the desire to explain the system of the world and the conviction that order and simplicity are proof of God's design. In his general scholium to the laws of motion and in his letters to Richard Bentley, Newton made explicit the hope that his discoveries would convince rational men that the Creator was at work and that this would, in turn, expand the limits of moral philosophy. Hutton shared this theological conviction, and *Theory of the Earth* is dedicated to the proof, through the phenomena of geology, that there existed a divine plan for the constant regeneration of the earth's productive capacity through the cycle of erosion and deposition.

Although he was trained as a physician Hutton actually was a gentleman farmer, and the source of his early interest in geology was his concern to understand how new soil might be created to replenish exhausted and overworked land. His conclusion was that the erosion cycle, superficially destructive, was the divine solution to the problem. The Lord, in His infinite wisdom, ordained that the land should be worn away and carried to the sea, where new land could be created and later raised above the waters. The gradual erosion of the rocks, endlessly repeated, created a constant soil bank; the system of the world was self-renewing.

When Hutton came to treat the question of why beds of loose material were not thrust up from the oceans instead of consolidated sediments, he argued the outlines of the divine plan; if God's purpose was a replenishment of the soil, there was no sense or economy of motive in such a phenomenon. Hutton's conviction in the matter was more particular than a simple advocacy of the erosion cycle as a general solution. Volcanoes existed, he said, because their expansive and uplifting power was needed to raise the strata; rivers which then cut out valleys and broke the rock to soil provided a means whereby plants and animals might find places in which to flourish (Hutton 1795:II, ch. 14).

This unrestrained deism fired Hutton's early opponent, the Irish mineralogist and chemist Richard Kirwan, who detected the pestiferous implications of Hutton's scheme of limitless time and avoided the contagion of such an unscriptural conception by reading as little as possible of what Hutton had written. Unimpeded by considerations of detail, Kirwan's attack was direct and broad. Bishop Ussher's conclusion that the earth was created on October 23, 4004 B.C. was not yet, in Ireland at least, a source of amusement for enlightened savants but rather a triumph of devout scholarship. The postulation of a time scale as vast as Hutton's was an abomination and a danger to revealed religion, and Hutton was an enemy to be dealt with severely. Hutton's second most formidable opponent, the émigré geologist Jean André Deluc, had the joint advantage that he knew something of geology and had scrupled actually to read Hutton. His strictures, thus informed, were no less severe, if more temperately worded: Deluc was also inspired to defend a geology compatible with Genesis.

When Playfair went to work on the logjams of Hutton's style, he sought to drag the sound timber from the thickets of natural theology and to present Hutton's ideas neatly cut and corded, ready to produce scientific light rather than theological heat. He suppressed Hutton's ubiquitous deistic teleology, leaving only a faint odor of sanctity hovering over the remains, and cloaked Hutton's own devotion to the divine plan in a style that demanded consideration of the substance of the theory before the suspected import.

But Playfair had another aim, which Hutton could not have achieved by any economy of style or theological caution—to explicate both Hutton's theory and his importance for geology. Playfair adopted the philosophy, typical of British thought in his time, that the progress of any science was a series of well-intentioned but fallible steps culminating in Newtonian perfection. Hutton's system could not pass away or be denied, because it stood in relation to previous theories of the earth as Newton's had stood in relation to those of Descartes and Aristotle. No theory in geology thus far, Playfair admitted, had long outlived the discovery of new facts. Yet, he argued, "it is, on this account, rash to conclude that in the revolutions of science, what has happened must continue to happen." Playfair realized as philosophy what Hutton had advanced as method: "When . . . the phenomena are greatly varied, the probability is, that among them, some of these *instantiae crucis* will be found, that exclude every hypothesis but one" (Playfair 1802: 512–513).

The real structure of the Huttonian system was more deductive than empirical and began with the conviction of the world's simplicity and order. A given phenomenon—an unconformable contact, the intrusion of granite into sedimentary beds—was not an isolated instance of the infinite possible combinations of mineral aggregates but a decisive clue for the sharp detective of a *determinate outcome* predictable within narrow theoretical limits, a clue to the real history of the world, astonishing in its simplicity and beauty.

Whereas Playfair was responsible for the distortion and reinterpretation of Hutton's thought, to Sir Charles Lyell we owe the standard historical account of Hutton's contemporaries and predecessors. Part of Lyell's celebrated *Principles of Geology* was devoted to a historical survey of geology from the tenth century (Lyell 1830–1833:I, chs. 3, 4). Lyell was not particularly generous to Hutton in this survey and criticized him for his ignorance of paleontology, his reliance on a theory of the earth's internal heat, and his catastrophic theory of continental upheaval. Hutton had written in

Theory of the Earth (in a passage expunged by Playfair) that "the theory of the earth that I would here illustrate is founded on the greatest catastrophes which can happen to the earth, that is [continents] being raised from the bottom of the sea and sunk again" (Hutton 1795:II, 124). To say that Hutton banished catastrophes from his *theory* is technically correct, for he refused to discuss them on the grounds that evidence was lacking. But it should also be clear that very few "catastrophists" in the history of geology ever invoked anything more violent than Hutton did himself. The cataclysms were for Hutton part of the history of the earth but not a part of his theory of it.

Lyell followed Playfair in allowing that Hutton had "labored to give fixed principles to geology as Newton had succeeded in doing to astronomy" but argued that Hutton had lacked the data to achieve this worthy aim (Lyell 1830–1833:I, 64). Lyell had definite and not very modest ideas about who the Newton of geology actually was—as the title *Principles of Geology* suggests. One would not be surprised to find that Lyell had made the analogy explicit and had called the work *Philosophiae Naturalis Principia Geologia*. The message was clear. And if, as W. H. Fitton suggested in 1839, Lyell had given insufficient credit to Hutton for the parts he had borrowed and Lyell's treatment of Hutton was unfair (Wilson 1972: 510–511), then Lyell's treatment of Hutton's opponents was grossly unjust and amounts to a propagandistic distortion of their views. Two recent writers have discussed Lyell's distortions in detail, analyzing their nature and the reasons for them (Rudwick 1970; Ospovat 1976).

Lyell was a great geological writer and traveler, and *Principles of Geology,* in twelve editions between 1830 and 1875, was an enormously influential book. Charles Darwin took it with him on the long voyage of the *Beagle* and credited Lyell with the inspiration for the vast time scale that figured in the theory of evolution. Retrospectively, this fact has focused attention on Lyell as a kind of torchbearer in a historical relay race—having laid the foundation for modern geology, he passed inspiration on to the founder of modern biology, connecting the cycle of revolutions.

Lyell's theory of the earth was in many respects similar to Hutton's but was even more wedded to the notion that the present is the key to the past; the nature of the agencies that change the earth's face has not changed. The forces at work are erosion and sedimentation, the slow rise and fall of the levels of land and sea, the action of earthquakes and volcanoes; there was no need, Lyell

argued again and again, for intermittent catastrophes and cataclysms to be invoked as explanations.

Lyell's version of the theory of the earth, which came to be
called uniformitarianism, was a philosophic extreme. Most other
writers allowed somewhat more variation in the nature and intensity of forces in past time. Henry De la Beche, for instance, urged
that periods of intense activity (geologically speaking, of course,
and with a notion of intensity measured across millions of years)
might alternate with periods in which the earth was quiescent. On
the other hand, Georges Cuvier argued for a series of catastrophic
deluges with successive extinctions of life forms as a joint explanation of the discontinuities in the fossil record and of the unconformable contacts in the sedimentary record.

In setting the stage for his polemical attacks on these opponents, loosely and tendentiously lumped together as "catastrophists,"
Lyell assailed an earlier school of geological thought—opposed to
Huttonian theory and also linked to catastrophism: neptunism.
Thus he could portray himself as successor to a victorious Huttonianism and his opponents as the unfortunate inheritors of a dis-
carded theoretical approach.

The neptunists were advocates of a theory of earth history in
which the activity of a primeval universal ocean formed the geological record as we see it. Viewed from the eighteenth century,
the theory had the theological advantage that it seemed to accord
with accounts of the biblical flood, while its geological rationality
stemmed from the evident fact, also present in the Huttonian
theory, that sedimentary beds were created and cemented at the
bottom of the sea.

Best known among the neptunists was the Saxon mineralogist
Abraham Gottlob Werner (1749–1817), whose views on the nature
of earth history were distorted by Lyell in order to discredit the
heritage of neptunism. Lyell did not obtain his information about
Werner's views from Werner's own work; his knowledge seems to
have been derived principally from a eulogy by Cuvier, itself a
mine of misinformation about Werner, and from the geological
works of Werner's students, such as *Traité de géognosie* (1819) by J.
F. d'Aubuisson des Voisins. His knowledge of the theories of Peter
Simon Pallas, another important German writer, also seems to have
come from Cuvier.[2]

2. The eulogy was delivered by Cuvier before the Académie des Sciences in
1819 and is available in English in Wm. Jardine, ed., *The Naturalist's Library* (London: Bohn, 1860), 29:19–40. Cuvier admired Werner, but his stories about Wer-

Lyell's exertions to discredit the neptunist school as a pernicious and retrograde influence on geology were concentrated in an attack on Werner, though there is much in Werner that is not neptunist and much in neptunism that owes nothing to Werner. The result of Lyell's attack was the creation of a legend around Werner that was the inverse of what Playfair had done for Hutton—in Lyell's account Werner becomes a counterrevolutionary, a nemesis (Lyell 1830–1833:I, 55–60; Ospovat 1976). The Werner legend (aptly named by Alexander Ospovat) has several layers. It begins with a hazy, incorrect, and schematic presentation of Werner's supposed views, includes a criticism of these views as harking back to theories of the biblical deluge, and finally attributes the system's faults to aspects of Werner's temperament, a psychological reduction later magnified by Archibald Geikie. While Lyell freely mixed Werner's own views with those of his students, he placed on Werner alone responsibility for the popularity and falsity of the neptunian theory. The resulting portrait of Werner and the Wernerian system looked like this: Werner developed a system of practical mineralogy far superior to any other of his time and taught that system at a mining academy in Freiberg, Saxony. Because of his excellent teaching, the academy became an international center of mineralogical science in the late eighteenth and early nineteenth centuries. Werner's solicitude and affection for students, his fine mineral collection and library, his small classes and individual supervision in the classroom and the field, the clear and systematic presentation of the material, his insistence on minute observation and careful mapping, his proclamation of a natural and empirical method, all served to win him universal acclaim and an international following. However, the legend continued, the greatness of Werner as a teacher of mineralogy blinded students to fatal flaws in his theory of mineral genesis and earth history. Werner's system, which argued that the strata had successively precipitated out of solution from an original ocean, was rigid and anachronistic. Werner's observations, with typical eighteenth-century parochialism, were confined to the neighborhood of Saxony (where his theories were confirmed) and the rest of the world was ignored. Against

---

ner's supposed peculiarities are the greatest source of the notion that the neptunian system was somehow the individual and strange product of a single man. Lyell was forced by his ignorance of the German language to rely on Cuvier's account of Werner's life. Archibald Geikie later followed Lyell's lead and even lifted generous extracts of Cuvier's memoir into chapter 7 of *The Founders of Geology*. For Lyell's account see *Principles of Geology* (1830), ch. 4, p. 54.

reason, observation, and established tradition, Werner argued that basalt was not a volcanic rock but a chemically precipitated sediment (Lyell 1830–1833:I, 58). He refused to see volcanoes as important in the life of the earth, denied that the earth had a great store of internal heat, and would not countenance the notion that erosion was of geological importance. Werner's students, swept away by his presentation of the system, excited by its apparent rigor and order, bound to its author by ties of personal affection and gratitude, were only later, and with great reluctance, forced to repudiate the system of their beloved master on the evidence of their own observations. Many of his students "converted" to the plutonist camp after seeing the volcanoes of the Auvergne region of France and the associated columnar basalts and lava flows, the very phenomena that had figured in the development of the theory of the igneous origin of basalt by Etienne Guettard and Nicholas Desmarest in the middle of the eighteenth century. By the 1820s all but the most diehard supporters, and certainly the best geologists of the Wernerian camp, had abandoned Werner's theory and embraced Hutton's. But, the tale concludes, even more damage to geological science from Werner's popularity was still to come: the battle between the neptunists and the plutonists was so vitriolic that it led to a reactionary swing away from any theory or speculation whatever, making geologists "inclined to skepticism even where the conclusions deducible from observed facts scarcely admitted reasonable doubt" (Lyell 1830–1833:I, 71).

The etiology of Werner's erroneous theory, beyond the supposed parochialism of his observations, was attached to his mental and emotional character. Werner traveled too little, it was said, and was temperamentally unable to meet the simplest demands of scientific correspondence or a growing literature. He was beset by such a passion for order that he made everything, especially his theories, too determinate and systematic. In his personal life he allegedly devoted much time to seating charts for dinners, to arranging his library and mineral specimens, and to feverish punctuality. As Geikie later suggested, he based his fame on his lectures and wisely avoided publishing the details of his system; had they been put down in print, severed from the master's personal charm and power, they would quickly have been discredited. His overall system, in its dogmatism, its tunnel vision, its ignorance, was "more suggestive of the damp dark of a mine than the contact with the fresh and open face of Nature" (Geikie 1905:238).

Geikie's attacks on Werner had a dual basis: a desire to celebrate Hutton and an uncritical acceptance of Lyell's and Cuvier's accounts as accurate portrayals of Werner's theory. Later, this history was generally accepted by nonpartisan geologists and historians who took Playfair, Lyell, and finally Geikie as accurate sources and absorbed the view that Hutton was a truly revolutionary thinker and Werner a representative of an older, less complete, idiosyncratic view of the world, against which the Huttonian theory eventually succeeded.

This picture is demonstrably incorrect and must be changed. First, Hutton has to be restored to his true historical context and reunited with his own theory. Second, the burlesqued portrait of Werner as an isolated eccentric must be rejected in order to see the outlines of the highly successful system he helped create and the tradition of natural history from which it grew. The heroic juxtaposition of new and old ignores the complexities of the very period that gave geology its modern form and aims, and it hides the emergence of fundamental tensions in geological thought that are pertinent even now.

Once the systems of Hutton and Werner are restored to their original contexts, their clash takes on a different meaning. Some areas of substantive agreement were much wider than one would imagine, but differences of scientific method, different views of the proper scope and aim of geological science and of the relationship between fact and theory, made it very difficult for the antagonists (or indeed the mediators) to achieve anything like consensus. When the last echoes of the original debate were fading in the 1820s, these basic disagreements persisted, allowing substantial elements of the Huttonian and Wernerian approaches not only to survive but to provide inspiration and direction for the rest of the nineteenth century.

Gordon Davies has examined the role of Hutton and his ideas in a work that provides a starting point for a revaluation of Hutton's originality and importance in the history of geology. In a study of the history of British geomorphology from the end of the sixteenth century to the last part of the nineteenth, Davies has reconnected Hutton to the tradition in which he worked and has explained the role of Playfair, Lyell, and Geikie in constructing around Hutton a legendary history that served their objectives and reinforced their own picture of the vocation of the earth sciences (Davies 1969).

Following Davies, we see that throughout the seventeenth century the role of running water in shaping the face of the land—

fluvial geomorphology—was generally studied and appreciated in terms of cycles of erosion and deposition. However, the triumphant biblical chronologies of Ussher and Newton, which gave the earth an age of only six thousand years, raised strong objections to the idea that such cycles operating at the rate observed could have any general importance in the history of the earth—there was simply not enough time. This led, on the one hand, to a complete discounting of their importance and, on the other, to theories that raised their action to periodic and cataclysmic intensity in great floods or deluges. Rather than abandon the mechanism, supporters of cataclysmic theory varied its rate of action in order to squeeze it into biblically allotted time.

In the late eighteenth century Hutton revived the theories of the *secular* (long-term) efficacy of erosion and of cycles of erosion and deposition, giving them new life by abandoning the restrictive biblical chronology and postulating infinite time in its place. "Hutton," Davies argues, "was not so much the precursor of the late nineteenth century school of fluvial geomorphology, as the scholar who blended the simple, uncritical fluvialism of the seventeenth century, with the teleology of the eighteenth century" (Davies 1969: 196). Hutton's time scale, like that of his opponent Deluc, was a product of his theology. The theory that he propounded was identical in almost every respect to that of Robert Hooke and was actually inferior in its account of the erosion of strata (Davies 1969:91). Each of Hutton's great theories—infinite time, cycles of erosion and deposition, and the secular power of those cycles—was a repetition of geological ideas widespread in the seventeenth century, and their combination was based on a theological deduction (Davies 1969:180–183).

Hutton appears as a revolutionary only against his immediate background, a stagnation in British geomorphic thought in the late eighteenth century. Hutton's "birth of geology" was, in Great Britain, a revival; moreover, it was a revival that was preceded on the continent of Europe by decades, in the work of Torbern Bergman, Desmarest, Guettard, Johann Lehmann, Peter Simon Pallas, Saussure, and Werner. Thus, Davies argues, "one must sympathize with Lyell's judgement that Hutton's writings show little advance upon that of Hooke and Steno," his seventeenth-century predecessors (Davies 1969:195). The idea that Hutton was ignored, as Playfair said, and was rescued by Playfair from the obscurity of his unreadable style is also a myth. Hutton had been discussed and dismissed, on grounds both geological and theological, by Deluc,

Kirwan, Desmarest, and Leopold von Buch, *before* the full version of his theory appeared in 1795 (Davies 1969:185–186). The version of Hutton and his ideas that Geikie put forward was the work of a eulogist, not a historian. The idea of Hutton as the director of a powerful research team in Edinburgh, fashioning a new scientific theory of the earth, is a product of biographical similarity and true admiration: Geikie too was an igneous geologist from Edinburgh and had received his secondary education in the same school that Hutton attended (Davies 1969:178–179).

Of course, the Huttonian theory cannot be discussed apart from its own mechanism—the power of the earth's internal heat in cementing, reforming, and upraising the strata. Davies acknowledges that Hutton deserves recognition for the "realisation that granite was of igneous origin" (Davies 1969:212). However, this award of recognition itself contains a myth—that the igneous origin of granite was generally accepted thereafter. H. H. Read mused: "The question of the origin of granite is perhaps the most lively of geological topics today—but we should remember that it always has been. About every twenty years or so the problem has been finally settled, and a sort of uneasy peace has broken out" (Read 1957: 169). Shortly after Hutton's time, an uneasy peace did break out, and the theory of the neptunists that granite was a precipitate from aqueous solution was universally rejected in favor of a theory that it crystallized from a melt. But Hutton did not originate igneous geology, nor was his theory influential in establishing it in Europe, where it had already been extensively developed by Italian and French geologists.

Such information serves to clear the ground only of Hutton's exaggerated originality and is not a judgment on the adequacy or power of the theory in his own version or in Playfair's refined and theologically muted form. It merely restores Hutton to the context —fluvial denudation theory and natural theology—in which he worked and indicates that the elements from which he fashioned the theory were not new and that his reputation has been enlarged by attributing to him the definitive solution to questions that remained unsettled.

Before the encounter of Hutton's theory with its historical opponent can be reasonably evaluated, the neptunist theory and Werner's version of it must be separated from the distorted history of Lyell and Geikie. Its actual context must be restored. This restoration has already largely been achieved in the work of Alexander Ospovat, Martin Rudwick, Leroy Page, and others who have compared Werner's teaching of his theory with the version of it offered in Lyell's history have shown that the two are very different (Ospovat 1976; Rudwick 1970; Page 1969).

The biographical details offered by Lyell, borrowed by him from Cuvier, have some truth. Werner did make the mining academy at Freiberg an internationally important institution through his excellent teaching of determinative mineralogy, and he also taught a theory of the earth in which the fundamental mechanism in the formation of strata was not igneous but aqueous and had at its core the idea of the precipitation of strata from solution and their fixation by chemical cementation. His students were loyal to him, and only gradually, and with some reluctance, did they abandon that part of his theory in the face of contrary evidence. By the 1820s the theory in the form in which Werner had offered it was dead, and the theory of igneous causes—the central role of the earth's internal heat in creation of rock—was generally recognized as correct. But the description of Werner's theory in terms of the aspects of it that were later rejected, particularly the vastly mistaken idea that basalt was an aqueous precipitate, explains neither why the theory was popular when it was originally proposed nor why Werner could later be viewed on the continent as a great geologist, in spite of the fundamental error. Two explanations have been given for Werner's enduring popularity: sentimental attachment to the teacher and the man, and the estimation of his mineralogy, shorn of the neptunist theory, as an "unchallenged contribution to science" (Eyles 1964:106–107). Both explanations are aspects of the truth, but the deeper answer lies in Werner's place in the development of geological theory and practice in Europe, which long after his death retained aims and methods that he had espoused and of which he was the popularizer but not the originator. Werner, like Hutton, had a past.

Werner's earliest work, the foundation of his fame, was a system of practical or determinative mineralogy that he developed while a student at Leipzig. He called his system an *Oryctognosie*, literally, "the study of things from the ground." He had translated the mineralogical system of J. K. Gehler, *De Characteristibus Fossilium Externis,* and intended to publish the translation as his first contribution to science. In Gehler's usage, the word *fossil* (*fossilium*) did not have its modern sense—petrified organic remains— but, as in Werner's term *Oryctognosie,* indicated only "things dug from the earth"—particularly minerals. At the urging of the naturalist J. E. Kapp, who knew Gehler was dissatisfied with his own system, Werner devised a new mineral system, published in 1774 as *Von den äusserlichen Kennzeichen der Fossilien* (Werner 1774/1962:x).

At a time when the foundations of modern mineralogical chemistry and mathematical crystallography were being laid, Werner took the investigation of the mineral world in the direction of traditional natural history—toward very fine discriminations of color, taste, texture, smell, and hardness while systematically ignoring chemical composition, crystal structure, specific gravity, fusibility, and other physical characteristics. This choice in his earliest work has been understood, or misunderstood, as characteristic of Werner's fatal attachment to outmoded ideas and has reinforced the judgment that his ultimate influence was either obstructive or destructive. But as Albert Carozzi has shown, Werner's choice was not atavistic or ignorant. Werner was aware that there were foundations for a mineralogical system other than external character, such as internal character, unique physical properties, or place and mode of occurrence. Nor did he see external character as the fundamental basis of classification—he agreed that proper natural classification could proceed only from chemical analysis (Werner 1774/1962:xxvi). In later years he was quite prepared to integrate his work with the chemistry of Berzelius (Anon. 1817:183).

His motive in choosing was pragmatic, and he offered not a natural mineralogical system but a systematic artificial method: "to classify minerals in a system," he wrote, "and to identify minerals from their exterior, are two different things" (Werner 1774/1962: xxv). He chose external character as a foundation for his system of identification because it was the best known, simplest, and most convenient method for the practicing mineralogist. He thought that practical mineralogy had been advanced by the study of mineral formation and distribution, but he argued that too often a statement of location, use, or composition, or a list of reputed principal characters had been assumed sufficient to distinguish minerals from one another and had then failed to be diagnostic. He "would rather have a mineral ill classified and well described," he said, "than well classified and ill described" (Werner 1774/1962: xxix). This judgment bespeaks his family background—three hundred years of Saxon miners—and his occupation as professor of mineralogy at a state mining academy and councillor of mines for Saxony. Wherever Werner departs from natural philosophy and natural history (as a scientific activity) he does so in the name of expedience. Perhaps this pledge of allegiance to the demands of practice was deliberate—the mineral system he proposed was certainly a factor in his appointment to the chair of mineralogy in Freiberg, where such devotion to practicality must have seemed appropriate. Later, when he was comfortably established, his ideas took a vastly more theoretical form.

Werner recognized that his system of approach to minerals was not natural, but he was adamant that the system was real in the sense that the characteristics he chose to discriminate were not arbitrary or selected aspects of the appearance of minerals but rather exhaustive determinants of their appearance as natural forms. In this sense the system was "natural"—it dealt with minerals as an emergent level of organization in nature, above the chemical. When a mineral was pounded to dust, it remained the same— but once dissolved by acid or otherwise chemically attacked, a mineral lost its natural character and ceased to exist as the original substance. The quibble was not scholastic; it was analogous to the argument that water, electrolytically separated into hydrogen and oxygen, descended from the molecular level to the atomic, and ceased to exist as water.

At Werner's level of analysis, the well-trained senses could function without the intervention of instrumental or experimental techniques and Werner maintained this belief until his death. His 1816 emendations of his personal copy of the system reiterated the defense of his neglect of crystallography and the physical properties of minerals: these approaches were fundamentally impractical—"they are too time consuming and tend to confuse the ideas" (Werner 1774/1962:54).

An example of how Werner's system worked may be of some interest. Its foundation was his personal collection of minerals at Freiberg which, supplemented by contributions from his globetrotting students, eventually amounted to about 100,000 samples of minerals, stones, and ores. As a young man, Werner is reputed to have exhibited a prodigious ability to identify the source of mineral samples, not just by region but by the individual mine. From this vast collection he selected certain minerals as standard samples for particular distinguishing characters, much as a paleontologist selects an index fossil.

Color was an important external character, and Werner defined more than fifty colors from the type minerals in his collection. "Flesh red" was one of ten discriminable reds in a system that recognized six browns, nine yellows, six greens, six blues, four blacks, six grays, and seven whites. The other reds he called morning, scarlet, blood, copper, carmine, crimson, peach-blossom, cherry, and brownish. Flesh red minerals were pale red, a combination of crimson red and yellowish white, and the type sample was the red china clay from Rochlitz. Another color, straw yellow, was described as a blending of sulfur yellow and a little reddish gray, defined by the type mineral—a yellow jasper from Lessa near Carlsbad in Bohemia.

Werner's was a quaint system, cumbersome and discursive, requiring a great deal of effort to learn, especially the retraining of the senses to distinguish the varieties of character and color. But combined with similar discriminations of other characters, it was, in his time, the most powerful tool yet devised for the identification of minerals in the field.

Once learned, the system was carried to South America by Alexander von Humboldt, to Denmark and Norway by Henrik Steffens, to America by Benjamin Silliman, to Scotland, by Robert Jameson and nearly everywhere by Leopold von Buch. Confronting a mineral that was snow white, massive, of uneven surface, externally barely gleaming, internally splendent, of common luster, composed of large even sheets breaking into rhombic fragments, transparent, soft, ringing slightly when struck, and meager and rather cold (though less so than talc) to the touch, one could know with assurance that one held a piece of "selenitic gypsum" (Werner 1774:114).

The system was consistent and complete, capable of including any new mineral discoveries; the unique, complete ensemble of external characters, whereby a mineral was identified as something new, also immediately defined its relationship to the rest of the system and specified the means by which it could be identified again and again, without the intervention of any technique, nomenclature, or analytical system. One can also see the way in which the system could be supplemented by chemical analysis and crystallography.

The function of Werner's collection as a standard for the system calls to mind an anecdote that illustrates how his concerns and attitudes toward mineralogical science have been transmuted into part of the "Werner legend." Henrik Steffens was a Norwegian mineralogist who studied with Werner and who, by his own account, was an intimate of major figures in the scientific movement known as Naturphilosophie, in which direction he extended Werner's system. Volume 4 of his autobiography, which filled ten volumes and more than 4,000 pages, begins in his twenty-fifth year, when he set out to become a student at Freiberg. Much of the legendary Werner, punctual to the point of pedantry, demanding fealty to his fixed views, dominating his students' fieldwork closely, is evident in the account of Steffens's experiences in Freiberg. At one point, Steffens was in the lecture hall when a box of mineral specimens was being passed from hand to hand. At a time when mineralogists lectured to hundreds, Werner divided his classes into small groups to provide individual instruction and to teach the recognition of minerals from hand samples (Raumer 1859:205).[3] While the specimen tray was being passed someone jostled it and nearly spilled the contents to the floor, at which point, according to Steffens, "Werner turned pale and could not speak . . . it was seven or eight minutes before Werner could command his voice." One might assimilate this anecdote with others concerning an eccentric who "had a fire no matter what time of year," "wore fur over his bowels," and was "crazy about his stones" (Steffens 1863:82–83). But considering the role of the specimens in his system, Werner's reaction might better be compared with that of the curator of the standard meterstick near Paris who comes upon an assistant about to employ it as a crowbar.

Werner inhabited a world that could be exhaustively under-stood as it appeared, given patience, attention, and effort. It was not Buffon's or Diderot's world of illimitable richness and variety, which could not be completely reduced to order. Werner argued and demonstrated that the mineral kingdom could be systemati-cally and completely ordered by the application of a method that pursued the universal constancy of mineral species to the limits of human perceptual ability. Insofar as he succeeded in impress-ing this system on his students, he validated this approach to geo-logical science by the route of natural history alone, without the intervention of basic sciences, experimentation, or mathematical analysis.

In 1780 Werner added to the curriculum at Freiberg a course of lectures on a subject that he called *Geognosie*. In his mineralogical system of 1774, he had divided the study of mineral science into three parts. The first was *Oryctognosie,* practical or determinative mineralogy. The second was *Geognosie,* which treated the occur-rence and formation of mineral bodies, or *Gebirgsarten*. The third, mineralogical geography, dealt with mineral distributions and types

3. Raumer attended Werner's lectures at the urging of Steffens, and he con-trasts Werner's calm empirical approach within the realm of practical experience to Steffen's "winged enthusiasm" (1859:203) and "Mystical approach" to mineralogy (1859:75).

of mountains, their constitution, and their mineral resources. This preliminary and pedagogically motivated division has been the source of considerable confusion, though its introduction was designed to promote Werner's contention that indicators such as location and mode of formation should be excluded from determinative mineralogy. The confusion results from Werner's modification of his own views on the wisdom of this organization and from the publication by his students of a series of unauthorized emendations of his basic works that reflect only in part the alterations he continued to make in the lectures he gave in Freiberg.

After the appearance of his *Von den äusserlichen Kennzeichen der Fossilien* in 1774, Werner published only two other works of theoretical importance. The first was a pamphlet outlining the material of the course of lectures on geognosy, twenty-four pages long, entitled *Kurze Klassifikation und Beschreibung der verschiedenen Gebirgsarten* [A short classification and description of the different mineral assemblages] (Werner 1787). The second was a larger treatise, the *Neue Theorie von der Entstehung der Gange, mit Anwendung auf den Bergbau, besonders den Freibergschen*, translated into English in 1809 as *New Theory of the Formation of Veins, with Its Application to the Art of Working Mines*. Many of Werner's students were foreigners, and in trying to make his works available in their own languages, they translated Werner's terminology in ways that reflected sometimes his changing views and sometimes their own interpretations of his meaning. Historical interpretation is complicated by the fact that the word *Gebirge*, which has the unequivocal meaning of "mountain" in modern German, was in Werner's time, according to Ospovat, sufficiently elastic to denote "mountain," "mountain range," "mineral assemblage," and "formation"—a consequence of the fact that hardrock mining was then scientifically explored as the succession of mineral formations in mountain ranges; thus the term had a wide variety of meanings in the technical literature, of which Werner's works form a part (Ospovat 1969:251).

The Vienna edition of Werner's 1774 *Von den äusserlichen Kennzeichen der Fossilien*, published in 1785, substitutes the term *Mineralogie* for *Oryctognosie*, and *Lehre von Gebirgen* for *Geognosie* (Werner 1785:12). Picardet's French translation of 1790 employs extensive emendations by Faust D'Elhuyar, one of Werner's students, and continues to call *Oryctognosie* by the name *Mineralogie*, yet translates *Geognosie* as *l'art des mines* (Werner 1790:sec. 3). Which of these changes represents a change in Werner's views, and which are adventitious? Similar confusion reigns elsewhere if we try to separate

Werner's ideas from those of his students. Thomas Weaver's edition of the original mineralogical system, published in Dublin in 1805, contains additions from Werner's lectures in 1791–1792 and new mineral names coined by Richard Kirwan, to whom the translation is dedicated. Charles Anderson's 1809 translation of the *Neue Theorie von der Entstehung der Gange* is supplemented by the ideas of J. F. d'Aubuisson des Voisins.

The most elusive term is *Geognosie*. Werner acknowledged the term *Geologie* but reserved it for speculative systematizing of the whole of knowledge of the earth. Geognosy was a "hard science" subject that in modern terms falls somewhere among petrology, lithologic stratigraphy, and historical geology. The term was borrowed by Werner from the work of the stratigrapher and mineralogist J. C. Fuchsel; Werner's later lectures in geognosy expanded the original content of the term to something like geology, in which speculative material of every kind might find a place.

The organization and naming of the various parts of the earth sciences in the eighteenth century (and the nineteenth) form a complicated story that reflects the variety of competing theories and the search for a rational classification, which often degenerated into mere logomachy. Werner, in separating geognosy (knowledge of the earth) from geology (discussion of the earth), tried to separate knowledge from speculation but later violated this division. Most of his followers and students, whether adherents of the neptunist position or not, tended to describe their work as geognosy, while Huttonians showed a preference for the term *geology*, ignoring the niceties of classical philology. This dual tendency often gives the mistaken impression that geognosy and geology were definably different studies and that somehow the terms may be taken to show the contents of a work and the party to which the author belonged. Some scientific journals—for example, the Swiss science review *Bibliothèque universelle des sciences et des arts de Genève*—several times refashioned their tables of contents, sometimes distinguishing contributions in geology from those in geognosy, sometimes not (Anon. 1816:314). In 1818, a geological treatise by Scipione Breislak appeared, a late attempt to reconcile the main points of the Huttonian and Wernerian systems. Breislak employed a threefold division of the earth sciences that began with mineralogy, the study of the materials making up the crust. It was followed by geognosy, also an observational science that studied the arrangement of these materials in the crust and their relief. Finally, geology brought everything together, on a foundation of physics and chemistry, into a full history of the earth (Breislak 1818:204). Breislak was attempting to bring the substantive contributions of earth historians, "geognosts," under the aegis of the program first espoused by Hutton—the generalization of results in terms of basic physical processes. As late as 1902 the term *geognosy* was still in use; Geikie's own *Textbook of Geology* included a section on it as a combination of mineralogy and petrology (Geikie 1902:9–10). Breislak's sense of the term seems to be the one most generally understood and used by European geologists in the nineteenth century and is quite close to Werner's original definition, supplemented by a mineralogical geography.

Thus considered, geognosy was a most conventional subject for a mining academy, and Werner's subject was not a new branch of science but a continuation of a long tradition of study of sedimentary formations. For centuries miners had investigated the succession of the rocks in the earth's crust, seeking to identify the characteristic association of certain sequences of rocks with desirable minerals. By the late seventeenth century, theories of regular stratigraphic succession had been developed by observers including such well-known figures as Robert Hooke and John Ray. Particularly notable was the Danish physician Niels Steensen (Nicolaus Steno, 1638–1687), who attended the Grand Duke Ferdinand II in Florence and was a member of the celebrated Academia del Cimento, one of the first of the great scientific societies of Europe. Steensen's work dealt with the nature of fossils, the origin of mountains, the stages of the erosion cycle, and the mechanisms of the deposition and deformation of strata.

By the middle of the eighteenth century there were stratigraphers active throughout Europe: Guettard, Desmarest, Antonio Moro, Giovanni Arduino, John Strachey, Buffon, and a host of others. Mountain ranges were extensively studied both because of their economic importance and because the exposure of the strata was greatest in elevated portions of the crust. Peter Simon Pallas (1741–1811) surveyed the Ural Mountains in Russia and the mountains of Siberia and found in them a repeated tripartite structure. The summits of the chain, along the long axis, were composed of crystalline rocks, and on their flanks sedimentary beds of calcareous rocks, fossiliferous limestones, marls, and shales were covered (and flanked) in turn by looser deposits containing more highly developed organic remains. Pallas called these rocks, respectively, the primary, secondary, and tertiary sequences. The division indicates the presumed order of deposition or creation according to the principles of superposition—younger beds upon older; that the crystalline rocks were called the primary sequence reflects the observation of geologists of the time that sediments were everywhere underlain by such massive crystalline rocks. The same scheme had been followed earlier in the century by the Tuscan miner, mineralogist, and metallurgist Giovanni Arduino (1713–1795) who, based on his own field experience, added a fourth classification: volcanic rocks (Geikie, 1905:195).

This general organizing scheme was perpetuated in Germany by Johann Gottlob Lehmann (d. 1767) and J. C. Füchsel (1722–1773), who advanced these studies of regular successions with detailed maps of specific mountain ranges. Lehmann's work was in the Harz and Erz mountains on the border of modern Czechoslovakia. In his *Versuch einer Geschichte von Floetz-Gebirgen* (1756), he argued that these mountains, like those observed by Pallas and Arduino, had a characteristic tripartite structure.

In these ranges, the core—the rocks exposed at the summits, along the axis of the range—was composed of crystalline rock. The core was covered by *Flötz-Gebirge* ("sedimentary formations"), which contained fossils, and above them (that is, laterally to the outside of the range) was a third sequence of rocks of looser material, not fully compacted. Lehmann assumed, as had Steno, Pallas, and Arduino, that the core rocks, now highest in relief, were the first deposited and were therefore originally the deepest.[4]

Fuchsel, working in the province of Thuringia, was a pioneer in the identification of stratigraphic formations by their fossil content, and from his work can be traced an emphasis on seeing the succession of rocks not as random isolated layers but as "formation suites," as he called them, composed of a number of layers in a predictable sequence. Thus the original tripartite sequence was already being explicated in the mid-eighteenth century as composed of smaller sequences that also seemed to follow a regular and continuous pattern (Adams 1938:217).

In addition to these works (and many others), Werner also had access to the system of classification that was proposed by the Swedish scientist Torbern Bergman (1735–1784) in a widely circulated

---

4. The principle of superposition, that the deeper the rock, the earlier it was laid down, was a commonsensical proposition urged independently by many observers: Lehmann, Steno, Werner, Hutton. Although often celebrated as one of geology's great "laws," it is not a "law" at all and only applies to rocks that have not been tectonically disturbed. Overreliance on the principle was a great failing of the early Wernerians.

volume that in 1769 was translated into German and English: *A Physical Description of the Globe*. Bergman's work was the immediate and direct foundation of the system that Werner began to teach in 1780 and was an interpretation of the prevailing fourfold classification in terms of a neptunist theory of successive worldwide precipitations around the earth's crystalline core. Within a few years of the beginning of his course on geognosy, Werner extracted the proposed sequence of rocks from Bergman's book and published it as the short pamphlet *Kurze Klassifikation*. This was not, clearly, an attempt to take credit from Bergmann; rather, it reflected Werner's early commitment to separating the schematic presentation of the succession of formations from the theory of their genesis; the original division of fact from theory, of geognosy from geology.

Werner's *Kurze Klassifikation* carried over into his teaching the prevailing theory of stratigraphic succession current in the late eighteenth century: at the bottom, the primordial, primitive, original rocks (*Uranfängliche*), succeeded by the sedimentary bed or layer (*Flötz*), the volcanic rocks (*Vulkanische*), and the alluvial layer (*Aufgeschwemmte*). These major subdivisions were supplemented by a more detailed internal succession that clearly reflects the work of his German predecessors and his own observations (Werner 1787). In 1796, Werner intercalated a new major subdivision, the transitional (*Ubergangsgebirge*), between the primitive and the sedimentary formations—a change incorporated in his lectures but characteristically not issued in a revised version of the original printed sequence (Werner 1774/1962).

Implied but not specified in the presentation was Werner's theory of the origin of the earth's crust, a theory that relied, like Bergman's, on the idea that many rocks were precipitates from a primeval universal ocean. This ocean, with the system of precipitates that it released, is the hallmark of the aqueous theory of the origin of the crust, the neptunist theory, which battled against plutonism and volcanism, the igneous theories that Werner opposed. Werner never published his own version of a neptunist theory, a reluctance that Geikie interpreted as fear that its absurdities would be revealed once its contents were severed from the persuasive lectures of the master. The truth, as in most of the Werner legend, lies elsewhere.

To begin with, Werner's own field experience validated the theory. The Erz Mountains, on which he relied extensively for evidence, are a product of the second great episode of mountain making in the history of Europe, the so-called Hercynian or Variscan orogeny that took place at the end of the Carboniferous period in the late Paleozoic era, about 300 million years ago. The area from which Werner drew his observations was remarkably free of volcanic activity during this period, when much of the tin for which the mountains were exploited was emplaced. This absence of volcanic activity certainly figured in Werner's refusal to see volcanism as an important geological force. Prior to the Hercynian orogeny, in which the mountains arose, a period of intense metamorphism gave the basement rocks (Werner's *Uranfängliche Gebirge*) a uniform crystalline appearance, obliterating, through heating and recrystallization, the original clastic, or fragmented, character of the sediments. In the following period, the Triassic (beginning about 225 million years ago), much of Europe was covered by thick beds of sandstone and carbonate rocks, the so-called Buntsandstein, Muschelkalk, and Keuper formations (Brinkemann 1960:10, 12, 60, 70). These gave not only Werner, but Lehmann and Füchsel before him, the impression that the stratigraphic history of the period was uniform over a wide area, and perhaps over the whole earth. Written accounts of major mountain ranges in other areas of Europe, from Russia to Italy, seemed to support the idea of universal formations of the type Werner had observed in Saxony. Thus, in his own presentation of the system, "universal formations" were restricted to the primary and the lower part of the sedimentary, which Werner later renamed the transitional, and these were the sole formations interpreted as universal chemical precipitates from a primitive ocean (Adams 1938:222).[5] All subsequent formations, the remainder of the sedimentary, the volcanic, and the alluvial, were partial formations covering only a part of the earth's surface and were produced by a complicated series of chemical formations, mechanical depositions, erosion, and the minor action of volcanic fire (Werner 1787).

The "onion-skin myth" of an earth created entirely by successive universal precipitations of rocks from chemical solution or suspension is, as Ospovat has noted, an utter invention when applied to Werner. Such a theory may be ascribed to Bergman, but Werner's theory was much more complicated; nothing in his own work and little in the work of his students supports a "pure" version of the neptunist theory as a succession of chemically produced beds, absolutely invariant over the surface of the earth. This invariable

5. Adams relies here on d'Aubuisson des Voisins, but the information seems compatible with Werner's own views on the subject, collected from a number of early sources.

stratigraphy that we see in Geikie, for instance, is a misconception that has been uncritically repeated since the time of Lyell (Ospovat 1976:192).

Werner's actual theory employed the device of a universal ocean to organize the phenomena of geognosy into a coherent pattern. The primitive, or crystalline, rocks were precipitated out first and covered the uneven core of the earth. The passage of so much material from solution caused a substantial drop in the level of the ocean and explained why the summits of mountains were generally composed of crystalline rocks: above the level of the re-maining ocean, they did not participate in the subsequent deposi-tional history of the earth. By the time the transitional rocks had precipitated, the level of the ocean had dropped enough to allow masses of the emergent crystalline core, and the transitional depos-ited on it, to erect barriers that divided the world into continental and oceanic areas and marked the end of the "universality" of the ocean.

Turbulence in these new oceans eroded the original formations and resulted in the laying down of mechanically produced beds together with further precipitations: these formations were re-gional or even local. Werner postulated great fluctuations in the level of the oceans (equivalent to a theory of oceanic transgres-sion and regression an enduring part of all subsequent theories) throughout the remainder of the sedimentary period to explain the diversified character and complex succession of sedimentary beds. In this latter period, Werner argued, most of the great features of relief were produced by further oceanic turbulence—valleys and cliffs were formed; slumping, slipping, and fracture took place in already deposited sediments, and these fractures were filled with new mineral precipitates, producing veins of ores. Progressively the process had slowed and become more regular until, in his own time, it was nearly complete.

In interpreting this record for his students, he had not estab-lished the exact position of many formations, and the full history was not worked out; in spite of his reputation for "fixed views," reported by Steffens, the system was open to revision at many points. Indeed, Werner's practical, professional commitment to mining, and the use of geognosy as a means for miners to locate valuable minerals contained in characteristic formations, made re-vision a necessity as well as a virtue. While Werner was certainly committed to a neptunist theory, and one that contained the seeds of its own demise, his goal above all was the empirical establish-

ment of regular successions of strata wherever they appeared and
the immediate employment of the knowledge of that succession to
serve practical and economic ends.

This motive, in the hands of Werner's own students, was the
principal means of the destruction of the neptunist theory, not the
real or imagined superiority of the Huttonian theory of the earth.
If the Wernerian system of geognosy had a particular disadvan-
tage, it did not consist in the fragility of an inner logic of special
delicacy, or a tendency to collapse under the weight of emendations and expansions. Rather, the system, as in the case of his mineralogy, was mostly *method,* and as it became larger, the systematic
aspects began to dissolve back into natural history—it became more
and more a history of the sequence of beds and less and less a
theory of the earth. This flexibility made the theory durable in the
fight of the British neptunists against the Huttonian school, which
was closely bound to a set of precepts that were abstract and theoretical and that could not, in consequence, be easily modified or
abandoned. In short, Werner appears not as an original thinker
and (in particular) not as the author of a narrow and inflexible
theory of the earth but rather as a systematist who brought together, generalized, and taught an approach to the study of the
earth that had great practical utility.

Once the rationality of Werner's system is appreciated, the need
to explain it away in terms of aspects of Werner's character disap-
pears. Werner was ill suited to controversy both by temperament
and circumstances. He was consumed by teaching, by supervision
of fieldwork, and above all by his duties as councillor of mines for
the kingdom of Saxony. The involved nature of his mineralogical
system made personal attention to students mandatory, and as his
fame grew, his responsibilities at the Bergakademie Freiberg increased proportionately. Because he used his students to extend
the mineralogical survey of Saxony, on which he spent more than
twenty years, he devoted great attention to the planning and evaluation of their fieldwork in the hundreds of miles of mine tunnels
around Freiberg, specifying for them particular routes and particular formations to be observed (Steffens 1863:84–85). This patient
attention was misinterpreted by a free-spirited philosopher like
Steffens as an attempt to dominate him by enforcing "fealty to
fixed views," as if Werner supervised his fieldwork to prevent Steffens from discovering the truth.

The volume of Werner's correspondence grew with his fame
and soon overwhelmed him. He did not like writing letters and

dreaded losing the time that scientific correspondence required (Anon. 1817:188). If he hated writing in general, as the legend specifies, he was apparently able to overcome this distaste long enough to turn out many thousands of pages of mining reports in the course of his official duties (Ospovat 1976:191). And so it goes with each of the anecdotes concerning his peculiarities. Feverish punctuality? A very busy man—part professor, part civil servant—with little time to waste. Fascination with seating charts at dinners as an expression of his mania for order? Perhaps, but the more likely cause was the volume of visitors he entertained from all over the world and upon whose conversation he depended to keep him in touch with events outside Freiberg. As for his passion to arrange mineral specimens and the books in his library, it seems dubious to treat a passion for order as a liability in a taxonomist, and in Werner's case it seems less than reasonable to criticize him for not reading all of the books he purchased to build up the university library in an institution that his own fame made a world center for the study of geology and mineralogy.

Werner and Hutton can best be seen as figures representing the culmination of traditions in the study of the earth that were already passing out of existence as their theories achieved a mature form. Hutton's combination of geology and natural theology would not long be countenanced in Britain and had already been abandoned (or very nearly) on the continent. Werner's use of external characteristics and his attachment to an artificial (if effective) systematics that ignored chemistry and crystal structure and was shaped by practical aims was insufficiently general for a realm of study aspiring to the title of science, and the form of the neptunist theory he embraced had not long to live in any case. Yet to strip these theories of their older elements, as earlier historians have done, and identify them merely by their most solid achievements and glaring errors, deprives both constructions of their rationality and their very reason for existence. Further, and more serious, it has made it very difficult to understand why the "neptunist-plutonist" debate of the first decade of the nineteenth century unfolded as it did and why the Huttonian theory did not emerge immediately triumphant, given its great similarity to modern geology and its superiority over the now nearly forgotten ideas of the neptunists.

CHAPTER TWO

The Convergence of Geognosy and Geology, 1802–1818

In the version of the neptunist theory proposed by Werner, there was very little role for heat as a geological agent. Strata, whether precipitated chemically or deposited mechanically after erosion of preexisting formations, were solidified and hardened by chemical cementation and crystal growth. Veins and seams cutting across such strata, particularly basalts, were the filling of fissures by aqueous solutions that subsequently solidified. Surface irregularity and the tilt of sedimentary beds were caused by the shape of the irregular core on which they were precipitated, as in the primitive and transitional formations, or by local slumping, slipping, and accommodation to stresses during the consolidation of the higher formations, particularly the sedimentary rocks.

The universal presence of weathered fragments, detritus, and alluvial material at the surface of the earth was seen by Werner and many of his followers as the result of a final epoch of oceanic turbulence in the history of the earth, after which the oceans retreated to their present basins. To counter the criticism that there was not enough room in the ocean basins for the remnants of the original universal ocean, it was later argued that much water had disappeared into the earth's interior with a violent finality or perhaps had been drawn off by a passing comet in this early period.

Hutton's theory was different in every major respect. All the strata were erosion products, and their consolidation was always a matter of fusion by the earth's internal heat while they lay deeply buried. This same heat was responsible for the folding and defor-

mation observed in sediments, whether during the period of consolidation or later during the uplift of the finished continent. The
veins and seams of basalt and of granite cutting across the sediments were igneous extrusions and intrusions—molten rock forcing
its way upward toward the surface, with occasional violence, as in
the case of volcanic eruptions. Loose material at the surface of the
earth was evidence of the ongoing cycle of erosion in the "Huttonian earth-machine," where creation and destruction of sedimentary material went on perpetually.

In both theories, all the major points were referred back to the
action of a fundamental agency: heat in the Huttonian and the
primeval oceans in the Wernerian. In the latter system, the appearance of the strata was related exclusively to the original conditions
of their creation and deposition. The cyclic nature of Huttonian
geological process had no place in geognosy, and the scale of infinite time that accommodated it was completely disposed of—or
rather was never considered. Hutton saw the present geological
record as a moment in a constantly repeated process, which would
eventually destroy without leaving a trace of the order that currently
prevailed. In the Wernerian scheme of the earth's past, on the
other hand, the present rocks were the whole history of the earth,
and the record was now substantially complete and inalterable.

This profound difference in outlook proved a fertile ground
for controversy, proponents of both sides dismissing the reasoned
deductions and crucial instances of the other from the standpoint
of elements that were basic to their own world view. Analysis of the
debate between opposing theorists, which took place principally
between 1800 and 1820, provides much insight into the nature of
geology and geognosy during the period precisely because there
was no determinate outcome that can be described as a victory for
one side or the other.

Neither Hutton nor Werner took part in the debate directly.
Werner was disinclined to controversy, publication, and correspondence, and he was too busy. Although Hutton's circumstances
would have freed him for a controversy with Werner, had he been
so inclined, he found his principal opponents among those geologists who would not adopt or condone his argument for intelligent
design of the universe. Chapter 3 of *Theory of the Earth* (Hutton
1795:I) began as an exposition of previous theories, but Hutton,
after managing to treat only Burnet, Maillet, Buffon, and Deluc,
trailed off into a polemic about design; railing against vulcanists
and neptunists alike, he never even mentioned Werner by name.

Playfair later noted that Hutton was unaware of contemporary theory, but this comment is belied by Hutton's many references to
major theories in the course of arguments that appear in the second
volume of *Theory of the Earth*. He was certainly *unwilling* in his
major work to give adequate accounts of the views of his contemporaries on general theoretical matters (Bailey 1967:74).

With the death of Hutton, defense of the theory passed to
Playfair, who published *Illustrations of the Huttonian Theory of the
Earth* in 1802. Immediately, the work was answered by an Edinburgh neptunist, John Murray, in *A Comparative View of the Huttonian and Neptunian Systems of Geology: In Answer to the Illustrations of
the Huttonian Theory of the Earth by Professor Playfair* (Murray 1802).
Murray undertook the reply because he found Playfair's account of
the Huttonian theory so compelling that it needed an immediate
answer to counter those of Hutton's doctrines that "whatever may
be their ingenuity and novelty, appear visionary and inconsistent
with the phenomena of geology" (Murray 1802:iii). Playfair's version of the Huttonian theory has since become esteemed as a scientific classic, while Murray's book has suffered the obscurity of the
rest of the Wernerian system and approach. Even so, Murray's is
without doubt the most reasonable and compelling presentation of
the Wernerian theory that has ever appeared. A summary of the
main points of the argument not only shows the contemporary
plausibility of the Wernerian account—and the weaknesses of the
Huttonian system from the Wernerian point of view—but high-
lights very well the contrast of geology and geognosy as they were
understood at the time.

Yet, it must be stressed that Murray took it upon himself to
place Werner's teachings in a general theoretical context as a way
of meeting Playfair point for point and in a way Werner might not
have condoned. It is particularly notable that Murray did not object
to Hutton's attempt to create a general theory of the earth, though
this was a persistent criticism of Huttonian geology by Wernerians
and by uncommitted observers of the two schemes. On the contrary, Murray's position was that "Systems . . . are in the sciences
what the passions are in the human mind: they may be the source
of great errors, but they are also the cause of great exertions . . .
objects apparently minute acquire an interest and importance;
views are suggested which often lead to real acquisitions; facts are
analyzed which would have remained isolated; and relations traced
which would not have been observed" (Murray 1802:v). For Murray, as for most contemporary Wernerians, the neptunist theory

was one of "strict inductions"; that is, it pertained to a description of the geological succession under the aegis of a mechanism of formation so plausible that it remained generally unexamined. Little attention was paid to the organizing generalization, and much attention to the detailed elaboration of the phenomena within its scope, an approach that characterized a number of subsequent theories, particularly that of the secular contraction of the earth.

However, Murray felt that driven by the inquiries of the theorist (Hutton), "it is perhaps necessary to complete the system; not merely to rest satisfied with the proof that minerals have been formed by solution, . . . but [to make] an attempt to show how this solution, and the subsequent consolidation from it, have been effected" (Murray 1802:12). This was, in a sense, a fatal step for Wernerian geognosy; Murray's move to the high ground of theory abandoned the empirical caution that surrounded Werner's own treatment of the neptunist hypothesis as an organizing generalization and locked Werner's geognostic succession into a theoretical and methodological context essentially foreign to it, but one which later "Wernerians" felt compelled to defend. This move, not quite complete, resulted in a tension evident throughout Murray's work between the safety of "strict inductions" and the satisfaction and polemical strength of a universal and adequate mechanism to explain the phenomena.

The main point at issue was whether the origin of strata was igneous or aqueous, and this was the question addressed by Murray. He began from the assumption that the world was once fluid. For the strata to contain impressions of animals and vegetables, he wrote, obviously the strata must be in a fluid state to admit them. The crystalline state of the strata and their existence in parallel beds also implied a former fluidity. Since the strata were once fluid and were now solid, the principal questions for geological theory were whether they had become fluid through fusion or by the action of some solvent and whether they had been consolidated by precipitation from solution or through igneous fusion. Thus setting the problem, he felt it would be an easy matter to find the answer by examining the phenomena in question (Murray 1802: 3–5).

Following these prolegomena, Murray gave summaries of the Huttonian and Wernerian views that are models of fairness. He summed up the Huttonian system as cyclic, igneous, one that refrained from an explanation of the formation of the globe, and one that assumed that its present state is not its original state (Murray

1802:12). His picture of the Wernerian system was a summary of
Werner's succession of primitive, transitional, and later sedimentary formations, with "the trivial additions of the products of volcanic fire, and the alluvial beds," and a short account of the formation of veins through the aqueous infilling of rents in the strata
during the period of consolidation, all of which implied that the
earth's present state reflects the conditions of its original creation.

In his attack on the Huttonian system, Murray concentrated on
its central doctrine of heat. He argued that Hutton was wrong a
priori, since any central heat would long ago have dissipated if the
earth was as old as Hutton inferred; the heat must have vanished
before the imperceptible work of destroying even one generation
of mountains could occur. By the time they were eroded away into
the oceans, no heat would be left to fuse the loose material into
strata of present hardness (Murray 1802:55). But there was a further and, from the modern standpoint, a more substantial objection to the system: no cause was given in the Huttonian system for
the expansive uplift that raised the strata from the bottom of the
ocean, nor were any reasons presented as to why it should not
cause havoc while acting, why it never elevated strata before they
were prepared (i.e., hardened), or why it always worked on the
bottom of the sea and never on already elevated land. Indeed, the
er.tire principle seemed "gratuitous and improbable" (Murray
1802:68–69). Of course, there had been a cause and a reason in the
Huttonian system, but the well-intentioned Playfair had removed
them in his drive to give Hutton a greater scientific plausibility.
Hutton had referred the nature and timing of the uplift to the
design of the Creator and then declined to speculate further on the
cataclysms. Thus the Huttonian doctrine that the continents had
gone *up* seemed speculative and incomplete compared to the
Wernerian view that the strata had been formed as the ocean went
down, and Murray was able to score an important point for the
plausibility of the theory he espoused, at the expense of the Huttonians.

When Murray turned to a defense of the Wernerian system on
its own ground, he went directly to the greatest objection raised
against it by opponents—it claimed the aqueous origin of many
substances that were known to be insoluble in water. This had been
Hutton's explicit ground for rejecting the (remote) possibility of
aqueous solution in the consolidation of the strata; many rocks
were cemented by substances like "fluor, sulpherous and bitumi-
nous substances, and feldspar," which were not water soluble. The

phenomena of chemistry, by means of which "we may be enabled to judge of that which is possible according to the laws of nature," dictated the superiority of the theory of igneous fusion (Hutton 1788:224). On the same ground, Geikie declared that the best argument against the Wernerian system was that it was a chemical system that knew no chemistry; the assertion that certain of its rocks were soluble in water was a manifest absurdity according to the chemistry of the time (Geikie 1905:217). V. A. Eyles, in a defense of the plausibility of the Wernerian system, apologized for Werner's doctrine of the precipitation of granitic primitive rock "in spite of the known insolubility of the silicate minerals composing them" by speculating that Werner had been led to the view by their crystalline appearance and knowledge of how crystals grow (Eyles 1964:106–107).

The point was a sore one, and in answering this attack, Murray was poignantly trapped by his own determination to flesh out the neptunist idea into a general theory. His first reaction was a momentary retreat into the safe confines of geognosy. We need not answer this objection, he proclaimed, for the present appearances of the phenomena are incommensurable with formation by fire, and we need not reason to the cause. It is the Huttonians who have taken the "high ground" of claiming a single cause and who must defend it everywhere it is invoked. The neptunists have stayed with "strict inductions" and may, in good scientific conscience, ignore the argument. Having done his duty to Werner's own strictures on geognosy as an empirical study, Murray then sallied forth on his own, true to his stated intention of meeting the Huttonians on their own ground, first by answering the objections raised considering the insolubility of certain minerals.

Of all the insoluble substances that gave the Wernerians and their system theoretical problems, the most insoluble was silex (silica), the major component of the primitive strata that were supposed by the Wernerians to have been formed entirely by chemical precipitation from the universal primitive ocean. Murray admitted that it was true: silex *was* insoluble in water. However, he reported, Dr. [Joseph] Black has found silex in solution in the exhalation of the geysers of Iceland and precipitated on the ground around them. Might not, then, he argued further, silex be soluble in a chemical soup with alkalis and later decomposed? If we restrict the Wernerian ocean to water, his argument ran, silex is not soluble, but if we assume a different character for that ocean, silex and all other presently insoluble minerals "though now insoluble from their state of aggregation, may still have had an aqueous origin." Murray then pictured the primitive ocean: very hot, filled with the "saline, earthy and metallic matters," enabled by its great heat to sustain great amounts of matter in solution and by virtue of its contents "to sustain a much higher temperature than pure water would" (Murray 1802:70–76).

But whence the heat? Murray appeared to be on the verge of an igneous theory of his own but immediately backed away. The heat, he said, was something from the earth's early history. The planet's original store of caloric was accumulated locally and was capable of producing the "greatest effects" at the surface. Moreover, "it has also been the opinion of several geologists (and as a hypothesis there is nothing to prevent it from being assumed) that at this period, the atmosphere was not formed," so that the "immense quantity of latent heat," now in the atmosphere, must have been contained in the fluid mass (Murray 1802:76). Thus Murray completed the separation of the systems from a theoretical viewpoint—Wernerianism or neptunism in his version was not only noncyclic, noneternal, and nonigneous but also nonuniformitarian. This statement does not mean that it was essentially a catastrophic system, though some events of great violence were attributed to the action of the universal ocean. Rather, the theory suggested that the geological column was created by a uniform process in the past, a process that no longer operates, or can operate, in the present scheme of things. For Murray and his neptunist colleagues, as for Hutton, "the present is the key to the past," but in a different sense. It is not that observation of geological processes now in action gives an exclusive key to the puzzle of the origin of rocks but rather that the presently visible record of the rocks is itself the key to their past genesis, which may be inferred from the careful study of their external characters—a study that reveals their unmistakable aqueous origins.

Murray's work helped carry the day for the Wernerians in Edinburgh, but in the trite but true phrase, in winning the battle he lost the war. His version of the primeval ocean could achieve any effect of crystallization and solution, alteration, disturbance, or deposition required by a neptunian theorist, simply by an alteration of its level, temperature, or composition. Seduced by the Huttonians onto the "high ground" of theory, Murray, too, claimed a single cause and was saddled with the consequent onerous responsibility of defending it wherever it was invoked. In doing so he exposed Wernerianism to the criticism, heretofore reserved for the Huttonians, that they relied on invisible speculative agencies. He showed the Huttonians where their own theory needed shoring up (some proof of internal heat) and cast all of his hopes on a chemical theory that others were able successfully to attack and demolish, though some years passed before this destruction was accomplished to universal satisfaction. Although Murray took a step that was ultimately fatal for neptunism, he succeeded for the time in preventing any resolution of the Huttonian-neptunist division through the avenue of theory and its elaboration. Claims on both sides would become more strident, but theorists would move no closer to a resolution of their fundamental differences.

Contests at the level of grand theory, however, were not the only way in which Huttonians and neptunists might meet. An obvious recourse to deciding the origin and cause of cementation of the strata was experimental test. Hutton and Werner were themselves resistant to the idea that laboratory testing could recreate or duplicate natural conditions. Werner objected because he believed that conditions during earlier ages of the earth were unknown and unknowable and experimentation was thus rendered moot. Hutton contended that it was impossible to recreate at the surface the combined conditions of temperature and pressure in the interior of the earth. In *Theory of the Earth* Hutton quoted with approval the opinion of the French geologist Guy S. Tancrède de Dolomieu (1750–1801) that application of heat to rock could not have the same effect upon the substances experimentally that it would under natural conditions because the time scale of application was very different (Hutton 1795:I, 66). Hutton was adamant on the subject and discouraged his friend James Hall from performing experimental investigations of the Huttonian theory; Hall complied with Hutton's wishes until after his death (Davies 1969:179).

Later, between 1805 and 1826, Hall succeeded in providing some support for the Huttonian theory of consolidation of strata by heat through a series of experimental simulations of conditions supposedly similar to those at the bottom of the ocean or deep in the earth. It had been one of the neptunist's charges against the Huttonian theory that limestone, one of the most important rocks in the crust, was undoubtedly of chemical origin because it quickly decomposed when exposed to heat; further, its chemical nature as a carbonate had been established by Hutton's friend Joseph Black. It was in response to this challenge that Hutton had first replied that heat and pressure, taken together, produced effects different from heat alone and that experimental tests were therefore inconclusive.

Hall undertook to test this notion by heating pulverized limestone under pressure in a plugged steel tube. At first he obtained indifferent results, owing to the difficulty of procuring gun barrels that would not expand and release all the pressure. In some of his trials, however, "chalk, or common limestone previously pulverized, was agglutinated into a stony mass, which required a smart blow of a hammer to break it" (Hall 1812:160). The problems of offering such evidence as support for a theory are evident in the reply of the neptunists: similarity is just that—it is not identity. At the most, they argued, Hall's experiment demonstrated that heat and pressure could *also* cement limestone, albeit in a way different from its natural mode of chemical consolidation under water (Sweet and Waterston 1967:84).

Another experiment by Hall addressed the crucial question of the igneous origin of granite and basalt. Previous experiments had shown the "result of the fusion of earthy substances, hitherto observed in our experiments, either in glass, or in a substance which possesses in some degree, the vitreous character" (Hall 1805:161). If Hutton's theory were true, it was difficult to see how granite and basalt had acquired their stony and crystalline appearance. Hall showed that when glass was allowed to cool slowly, it took on a stony, crystalline appearance but could then be reheated and cooled quickly back to a vitreous state. He performed similar experiments with lava, producing either glass or stone from the same material, depending on the speed of cooling.

Here again the neptunists presented opposition, particularly from Robert Jameson, who performed his own experiments on glass and concluded that Hall's experiments, far from supporting Hutton's theory, merely meant that at different degrees of heat, alkaline substances were subtracted from the melt, and this led to a change in the physical character as well (Sweet and Waterston 1967:87).

In a third series of experiments, Hall addressed the very substantial objection to the Huttonian theory of consolidation by heat alone: no amount of heat applied to loose sand, gravel, or shingle, he said, "would occasion the parts to consolidate into compact stone" (Hall 1826).[1] Hall was eventually able to make sandstones of

1. See also Rachel Laudan, "The Problem of Consolidation in the Huttonian System," *Lychnos*, 1978.

various degrees of durability by the use of salt as a "flux." In this simulation, Hall produced the sandstone by covering a layer of sand with a salty brine and heating the two together in a great iron pot from below, attempting to show that the process could take place at the bottom of the ocean, according to the Huttonian plan. In this case, in spite of the successful production of sandstone, the neptunists could quite rightly argue that the results supported their theory more than Hutton's, since the chemical nature of the union of particles was demonstrated by the necessary presence of a salt flux—against Hutton's own objections. Again, however, neptunists were forced increasingly to raise the temperature of the universal ocean to remain consistent, while the Huttonians were simultaneously forced to acknowledge a greater role for chemical consolidation and even aqueous origin of certain strata. By the time Hall had finished his experiments, the neptunian ocean of the Edinburgh Wernerians had become a hot soup indeed. By the teens and early twenties of the nineteenth century, the remaining neptunists espoused a sort of igneo-aqueous theory that tried to consolidate the best of Huttonian and Wernerian arguments into a compromise framework. Characteristic of these attempts were the theories of J. F. d'Aubuisson des Voisins and of Scipione Breislak (Breislak 1818; d'Aubuisson des Voisins 1819). It is interesting to speculate whether the later and dominant German theory of magma (a subcrustal undifferentiated melt that produced various volcanic and plutonic rocks by fractional crystallization under different conditions of temperature and pressure) might be interpreted as an evolutionary development of these igneo-aqueous theories— merely moving the hot mineralogical soup of the neptunists from the surface to the interior. Werner's own version of the theory had postulated an eventual displacement of the hot primeval ocean to a home below the crust.[2]

Be that as it may, experimental simulations could not provide confirmation or refutation for the competing theories. Hall's experimental evidence that heat *could* have been the agency by which strata were hardened did not amount to a convincing demonstration. Either because of contrary results or contrary interpretations of results, or because of the inability of parties to agree on the import of laboratory recreation of the behavior of substances in

2. Such a speculation could go a long way toward explaining how an ardent Wernerian like Buch could have convinced himself that mountains were born in the upheaval of crystalline rock in a molten state from below the crust. The idea certainly deserves future research.

their natural state, Hall's experimental defense of Hutton's theory proved inconclusive.

Since both sides had been able to erect plausible general theories, and experimental work failed to provide a resolution, the debate raged on. Some years later, in *Principles of Geology*, Charles Lyell asserted that the formation of the Geological Society of London in 1807 was, in part, a response to the excesses of the volcanist-neptunist, or "Edinburgh-Freiberg," debate (Lyell 1830–1833:I, 71–72). Lyell's juxtaposition of Edinburgh and Freiberg was tendentious and incorrect, for the debate against the Huttonian theory was carried on locally, by Edinburgh neptunists like Jameson and Murray, and employed arguments neither developed nor advanced by Werner. But Lyell was at least correct that many wished to abandon the debate over systems and return to field geology in deciding the matter of the igneous or aqueous origins of strata. This schism meant not a total withdrawal from comparison and testing of the Huttonian and neptunist theories but only a refusal to become involved in the attempt fully to establish either theory as all-embracing and true, after the manner of Playfair and Murray. An example of the more cautiously empirical approach can be seen in the work of George Bellas Greenough (1778–1856), first president of the Geological Society of London.

In 1805, armed with Jameson's *Mineralogy of the Scottish Isles* (1800), a Wernerian work, and Playfair's *Illustrations of the Huttonian Theory*, he toured Scotland to see some of the famous "crucial instances" offered in support of the Huttonian theory and to determine whether an impartial observer could succeed in resolving the major issue of igneous or aqueous origin of certain rocks (Rudwick 1962).[3] It was Playfair who had suggested the idea of crucial instances, a consequence of his Newtonian approach to scientific theory; he argued that among any set of phenomena some must stand forth as sufficient to eliminate one and support another theory.

Greenough visited various sites to look at the evidence for the two theories and was swayed by what he saw to accept the erosion cycle and the time scale from Hutton's work. However, proof of igneous origins eluded him. At Glen Tilt, where Hutton had exulted

3. Greenough's tour was a "general" one—political, economic, geological—a type of journey common at the time. The geological works of Leopold von Buch were famous as travel literature; his *Travels through Germany and Italy* and the record of his Scandinavian expeditions enjoyed popularity as travel literature and adventure—and incidentally gave his (and Werner's) views a wider audience.

over the clear evidence that granite veins had cut across the schist, demonstrating that granite was igneous and not a primordial precipitate, Greenough found the evidence inconclusive. Everywhere he went, Greenough looked for evidence of contact metamorphism —signs that rocks had been transformed by the heat of molten "whin," or basalt. In many areas where Hutton had found it, Greenough was not convinced by the appearance of the contiguous rocks that they had been so transformed. He did observe a coal seam that had been changed to coke from its contact with "whinstone", and he thought this a strong argument for Hutton's case (Rudwick 1962:119–120).

However, we should note that a burned coal seam was explained in the Wernerian system by a simple observation: the coal had burned. That it should be in contact with whin or any other rock simply showed the local succession of the formation and was not evidence that basalt had ever been anything but the aqueous precipitate it was supposed to be.

Greenough seems not to have understood Hutton's theory of consolidation of rocks under heat and pressure (Rudwick 1962: 119). Some of Greenough's notations do not bespeak a very good grasp of the Wernerian theory of aqueous origins, either. If Werner was right about the origin of veins (that they represented fissures in sediments filled later by mineral solutions), Greenough wondered, why were there sills? The very question indicates a thorough confusion about the nature of basalt in the Wernerian system. A basalt "sill" is easily explained in that system, though it gave the plutonists problems. For Wernerians, basalt was an aqueous precipitate, and to find it conformably interbedded with other sediments was the natural mode of its occurrence, requiring no explanation at all. Wherever the neptunists found them, basalt sills seemed evidence that conformed with their theory (Geikie 1905:259). The plutonists had the difficult task: to explain how the molten matter had been laterally injected for great distances between otherwise undisturbed beds.

Greenough's confusion is instructive. In the Huttonian system, basalt sills (conformably bedded) and dikes (cutting through successive layers) signify the intrusion of molten material, an identity of mode of emplacement. Everywhere basalt occurs, igneous intrusion is inferred to have taken place. The same is true of the Wernerian system: wherever basalt occurs, aqueous origin by precipitation may be assumed to have occurred. There is an important difference, however. In the Huttonian system, whether intruded as a dike or a sill, basalt is something that was emplaced *after* the sediments were laid down—as the very term *intrusion* implies. In the Wernerian system, on the other hand, basalt might occur variously as a regular part of the process of deposition, or as a subsequent infilling of a rent cutting across the regular succession. To puzzle out how the Wernerian theory of veins could account for a sill (horizontal basalt) is to make a Huttonian assumption about the time sequence of events—that basalt is always emplaced after the regular succession has been completed—and to apply it to the Wernerian structure, where it contradicts the whole scheme of things. If, as in the Huttonian theory, basalts were rocks that always interrupted the normal sequence, then the Wernerians with their notion of veins filled from above would be in a predicament to explain how a "vein" of basalt came to be horizontally interposed between other beds. How could such a vein have gotten there? Clearly, however, in the Wernerian scheme, a conformably bedded basalt has nothing whatever to do with the theory of veins, nor is the presence of a basalt, per se, an indication of something that happened later than the period of deposition of sediment.

Greenough's difficulties, both observational and conceptual, highlight the perils and frustrations of the scientist who takes his stand as a "pure empiricist." He wants the phenomena themselves to provide the answer without mediation. Yet in his approach there is at once too little theory and too much. Too little of a Huttonian to be persuaded by what he saw at Glen Tilt and Salisbury Crags, he employed a Huttonian assumption about basalt to "test" the Wernerian theory of veins.

The problem of resolving theory by reference to crucial instances was not merely a Huttonian strategy, nor were the Wernerians reticent to point to conclusive field evidence in favor of their own theory—yet conversions on either side failed to materialize. In 1814 the neptunist Giovanni Brocchi (1772–1826) lamented:

> When one shows them (the vulcanists) that the rocks they call lavas alternate at Aix fifteen or twenty times with calcareous strata, when one shows them that in several places they are full of marine shells which are found in the most perfect state of preservation, when one points to basalts emplaced upon a substance as combustible as coal [*carbon fossile*] as has been verified on Mt. Meissner in Hesse, and when one finally considers that these striking facts have made not the least impression upon them and that they follow their own course with indifference, one is forced to conclude that such changes of mind are extremely difficult; so difficult that one despairs of them. [Meunier 1911:348]

This problem is further illustrated in two celebrated anecdotes from the autobiography of Darwin. The first is a product of Darwin's student days at Edinburgh, where the neptunist Robert Jameson was professor of natural history. The only effect of Jameson's lectures on him was to form a determination never to read a book on geology or study the science. Darwin recalled: "I . . . heard Professor Jameson, in a field lecture at Salisbury Crags, discoursing on a trap dyke, with amygdaloidal margins and the strata indureated on each side, with volcanic rocks all around us, and say that it was a fissure filled with sediment from above, adding with a sneer that there were men who maintained that it had been injected from beneath in a molten condition. When I think of that lecture I do not wonder that I determined never to attend to geology" (Darwin 1882/1958:53).

Jameson (1774–1854) was one of the most ardent of the Scottish Wernerians and so imbued with the theory that, as L. A. Necker de Saussure observed in his *Voyage en Ecosse et aux Iles Hebrides* (1821), his "too scrupulous" adherence to Werner hurt the excellence of his observations. It is not too remarkable that Darwin should have been disgusted by his extreme partisanship.

However, this anecdote of Darwin's revulsion at the tyranny of theory, which has already been applied to this subject by Adams (1938), is accompanied by another that shows as well the dangers of the opposite extreme. In 1831 Darwin accompanied Adam Sedgwick to Wales on a geological excursion.

> On this tour I had a striking instance how easy it is to overlook phenomena, however conspicuous, before they have been observed by anyone. We spent many hours in Cwm Idwal, examining all the rocks with extreme care as Sedgwick was anxious to find fossils in them; but neither of us saw a trace of the wonderful glacial phenomena all around us; we did not notice the plainly scored rocks, the perched boulders, the lateral and terminal moraines. Yet these phenomena are so distinct that . . . a house burned down by fire did not tell its story more plainly than did this valley. [Darwin 1882/1958:70]

Darwin's phrase "how easy it is to overlook phenomena . . . before they have been observed by anyone" contains the key to the problem of deciding theoretical questions on the basis of observations. Darwin did observe glacial phenomena in the colloquial sense, but he did not see them as elements joined together by a theory and therefore did not remark upon them. What he lacked to understand them on that first visit was not powers of observation but

some concept of them that would have extracted an organized
body of fact from a jumble of stones. Darwin's experiences, like
Greenough's and Brocchi's, reveal that insufficient respect for the
extent to which theory guides observation leads to two undesirable
outcomes: one in which theory overmasters the phenomena and
blinds the observer to contradictory possibilities (as in the case of
Jameson at Salisbury Crags); and another in which, for lack of an
organizing generalization (a theory of glaciation), phenomena pass
unseen and are, from the scientific point of view, invisible and lost
to science until they are theoretically organized so they can be *seen*.
This necessity for a theory to guide observation was beginning to
be recognized by geologists at the time of Playfair and Murray and
was underlined within a decade in a strong statement, again by
Brocchi:

> Nothing is more common than to hear people rail against systems,
> and to see those common-place remarks brought forward, which
> are usually resorted to on such occasions—That the number of
> well ascertained facts is yet very limited; that it is impossible to
> establish any general axioms. . . . These remarks may, within
> certain limits, be all very true; but it is no less true, that many
> persons allow themselves to be deceived, by laying down princi-
> ples such as these; and, while they are exclaiming against the abuse
> of hypotheses, they seem to be ignorant of the use of them. My
> own opinion is, that had it not been for geological systems, the
> knowledge we should now possess of the structure of the globe
> would be scanty indeed; and that to those more or less ingenious
> theories, such at least as have not been mere speculations, we owe
> in great measure that accumulation of facts which may be said to
> constitute the true capital of science. Many of those details, re-
> specting the nature and the differences of rocks, their reciprocal
> connections, the order of their superposition &c. would have es-
> caped observation, or would have been passed over as indifferent,
> had they not been considered as possessing a peculiar value in the
> defence or refutation of some particular system. [Brocchi 1814:
> 16]

Brocchi's assertion of the desirability of a theory to guide research also recognizes the positive value of controversy in clarifying
and improving the store of true scientific capital. The debate between Huttonians and neptunists caused neither side to yield but
did force clarification of important methodological issues: the possibility of two theories, each with some claim to adequacy over the
same range of data; the suggestive but inconclusive nature of experimental data in deciding theoretical questions; and some recog-

nition of an organic connection between the theory and the observed data. Furthermore, the debate actually forced the theories to converge over a wide range of issues. In their developed form, both theories agreed that strata were formed at the bottom of the ocean. Both Huttonians and neptunists agreed that the fundamental basis of mineralogy and petrology was chemical composition. One of the Wernerian's strong arguments in the controversy over the nature of basalt was that since the "basalts" of Scotland and Europe were chemically different, field evidence concerning the igneous origin of Scottish whin could not be applied to the European basalts, however similar the external character. Even Hutton admitted, on the evidence of the Swedish mineralogist Axel Cronstedt (1722–1765), that the rocks were different (Bailey 1967:49).[4]

Both theories agreed that the present rocks were created in a complicated succession of events, involving heat, chemical action, erosion, and dislocation. While the Wernerians held for certain universal formations, they readily acknowledged local variations in succession. Both theories concurred on the periodic fluctuation of the relative levels of land and sea—the Huttonians urging the rising of the land, and the Wernerians the fall in the level of the sea. These alternatives, incidentally, remained theoretically viable to the first decades of the twentieth century.

Each theory was seen by its proponents to be drawn from empirical observation, each was dubious about the value of experiment, each condemned the other for its biased inattention to the true order of things and called for a resolution of differences by appeal to geological evidence. Each system held itself open to revision on the basis of new information and evolved away from a too great insistence on a single fundamental cause—the Huttonians made room for chemical cementation, the Wernerians allowed an increasingly important role for the earth's store of heat.

The fundamental differences were not resolved, however, and support for the aqueous origin of strata slowly declined, while the idea of erosion cycles gained ground. There was also increased acceptance of the importance of volcanic action and of plutonic intrusion and uplift. Yet these developments, in Europe at least, were not in any sense a victory for the Huttonian theory. There was no consensus on the time scale of geological action or the overall importance of erosion. Moreover, in spite of the ebb of support for

4. Bailey points out that two different provinces of intrusive rocks, one "Northern" and one "Southern," characterized by different alkalinity, were recognized until 1892.

the neptunist theory, Werner's reputation remained everywhere intact.

This is confirmed in a number of histories of geology written in the 1820s and afterward, when neptunism was "overthrown"; geologists persisted in naming Werner as a founder of geology and had little good to say about Hutton. In Cuvier's estimation, Werner completed the work begun by Pallas and Saussure, the transition from fantastical theories of the earth's origin and history to detailed observation and positive knowledge (Cuvier 1819/1860:19). Other geologists, who adhered to the idea of uniform rates of change, and who took the volcanist side on questions like the origin of basalt, revered Werner as a great geological master. Necker saw Werner and H. B. de Saussure as the two great founders of modern geological thought, even though he agreed with Hutton's interpretations of intrusive granite at Glen Tilt and had criticized Jameson for too slavish adherence to Werner's doctrines (Necker de Saussure 1824). Seeing Werner's views as representative of the mainstream of geology, he noted that the remarkable observations of Hutton, Playfair, and Hall were either unknown or so associated with an inadmissible theory of the earth that they made but a "feeble impression" on most geologists. Significantly, he added that the decline of the neptunist theory and its fall as a system had been brought about by Wernerian geologists—Buch, Humboldt, and François Beudant—and not by adherents of the Huttonian theory. The actual establishment of the theory of igneous causes in Europe, over the older neptunist doctrine, took a much different historical course from that suggested in the traditional historiography in which Huttonianism "triumphed."

In that history, as suggested in Chapter 1, the displacement of Werner's ideas was a reluctant step taken by his students on the basis of field observation. Particularly important were the conversions to plutonism of Leopold von Buch (1774–1853) and J. F. d'Aubuisson des Voisins (1769–1841). These ardent supporters of Werner's theory of the aqueous origin of basalt traveled to visit the extinct volcanoes of the Auvergne region of France, where Desmarest many years before had shown that the association of columnar basalts with lava flows was a strong argument for an igneous formation of basalt. Buch and d'Aubuisson des Voisins are supposed to have become convinced that Werner was wrong but to have concealed their "conversion" from Werner out of respect for the master. This interpretation of the failure of converted Wernerians immediately to publish their change of mind was first suggested by the French geologist Ami Boué (Eyles 1964:113).

As W. Nieuwenkamp has recently shown, the idea that Buch converted or recanted after his 1802 visit to Auvergne is a "facile invention" belied by Buch's own journals. Only after his visit to the Canary Islands in 1815 and subsequent trips to Italy in the 1820s did Buch convert to an igneous theory. When he did so, it was to a catastrophic theory of mountain uplift, the theory of "craters of elevation," vastly different from previous plutonic or volcanic theories and repugned uniformly by Hutton's Lyellian descendants (Nieuwenkamp 1970:552). After the original visit to the Auvergne volcanoes, Buch had argued that their igneous character could be explained as local melting and extrusion of buried basalts of aqueous origin. While this sounds like a desperate ad hoc stratagem, Kenneth Taylor has shown that this was Desmarest's original interpretation of the phenomena he first described in the Auvergne. Desmarest, in fact, was a convinced neptunist, who opposed the idea that the primitive rocks could have been formed by heat. Moreover, he argued that volcanoes were not a part of the uniform course of nature and were mere accidents, resulting from the melting of aqueous basalts by accidental combustion of subterranean coal seams (Taylor 1969). Werner could not have "overthrown" the idea of igneous basalt, as Lyell and Geikie asserted, because his theory of volcanism was substantially the same as Desmarest's—combustion of subterranean coal (Zittel 1901:60).

The conversion of d'Aubuisson des Voisins was as ephemeral. Geikie marveled that when *The Basalts of Saxony* appeared in English in 1814, the translator took no note of the author's conversion to the igneous theory in the interim period—the original work had been published in 1803 (Geikie 1905:244). The error is Geikie's. The only admission made by d'Aubuisson before 1814 was that the Auvergne basalts were volcanic in origin, as Buch had already acknowledged. That the basalts of Saxony and the Auvergne were identical was a judgment that did not appear until the publication of his *Traité de géognosie* in 1819. While this was indeed after Werner's death, it was also after Pierre Louis Cordier's (1777–1861) highly celebrated experimental determinations of the actual composition of basalt in 1816.

On the European continent, the theory of igneous origins had nothing to do with the Huttonian theory of the earth and was developed through the experimental researches of continental European scientists. Cordier, who resolved the basalt dispute with chemical and microscopic examinations, attributed the eventual victory of the igneous theory to observations by Lagrange and Dolomieu. The crucial data were observations on the earth's temperature gradient, and the original study of the phenomena had been carried through by Leibniz, Halley, Descartes, Buffon, and others, who adopted the ancient notion of a hot earth from considerations of the earth's figure and from experiment (Cordier 1828). This view had gone into decline in the eighteenth century in favor of the ideas of Werner, Pallas, and Saussure, who had argued that the radiant heat of the sun, the primitive store of caloric in the primeval oceans, and volcanic combustions were the important sources of heat; while many observers remained uncommitted, a plurality supported the views of Werner and Saussure. The overthrow of their long-dominant system came through the observations of Lagrange and Dolomieu. Cordier acknowledged that Hutton and Playfair, "in spite of the obscurities in which they enveloped their opinions and the physical errors they fell into while applying them to geology," managed to convince most of the British of the importance of the earth's internal heat, but only by citing the results of French scientists: for Cordier, Hutton was a passive and parochial advocate of a theory in which the crucial data were provided by experiments and observations conducted in France (Cordier 1828: 86–90).

For the rest of the century, European observers insisted on the positive influence of Werner, and the minor role of Huttonian theory. When Joseph Fournet (1801–1869) wrote a history of alpine geology in 1849, he named the first epoch the "Wernerian Period," characterized by the work of Buch, Humboldt, J. C. Escher, H. B. de Saussure, Dolomieu, and others. The second period, succeeding the Wernerian, was not associated with Huttonian insights but with the adoption of later English classifications of sedimentary terrains by Wernerian successors: Buch, Necker de Saussure, and Léonce Elie de Beaumont (Fournet 1850:243–245). Although Hutton had devoted a great deal of attention to the Alps in the discussion of his theory, he was not acknowledged to have made any contribution to their study.

In these historical accounts, Werner was not associated with neptunism but was credited with a methodical approach to geology and with the outstanding characteristic of empirical caution. Necker de Saussure inferred that the neptunist system was, in a sense, a separable metaphysics within the Wernerian approach, whereas the views of Hutton were inseparably bound to an unacceptable system (Necker de Saussure 1824). Cordier argued that Wernerians remained uncommitted to the notion of a hot earth for lack of empirical evidence, whereas the Huttonians and Hutton himself plunged

ahead from theoretical considerations alone. Finally, Fournet identified Wernerian theory as the first coherent account of stratigraphical succession; it was replaced not by any igneous theory of the
earth but by an improved theory of a similar kind—a better stratigraphical nomenclature.

While Werner's system contained substantial errors—aqueous
precipitation of granite, the denial of the importance of subterranean heat, a too confident approach to the regularity of stratigraphic succession—these errors were not overcome by subscription
to Huttonian alternatives, but by patient fieldwork, minute attention to particulars, and careful mapping of sequences, performed
by Werner's students according to the principles they learned at
Freiberg. When, as in the case of Buch, Necker de Saussure, and
d'Aubuisson des Voisins, Werner's students abandoned cardinal
tenets of the neptunian theory of the origin of strata, they nevertheless remained, in their own minds, Wernerians out of the scientific conviction that it was Werner's approach to the study of the
earth that had enabled them to revise their opinions away from
Werner's own and to come to a correct answer. Moreover, it is
instructive to examine the major portions of Werner's system that
were rejected. In each case Werner's neptunist deductions from his
geognostic observations—the idea of primitive crystalline rocks, the
series of chemical precipitates, and uniform successions over very
wide areas—were overgeneralizations from essentially correct observations. The presence of granite as an unconformable base in
almost all exposures he observed led Werner to believe in its primitive status. The idea of chemical origin of the strata was expanded
to include sediments and igneous rocks alike, and the schematism of
the universal formations was little more than a heavyhanded application of the principle of superposition.

One might legitimately, then, remain a Wernerian by revising
and limiting the generality of the phenomena without throwing
them out, and this is what happened to Werner's European students. Werner's approach had a certain desirable elasticity and
openness to revision, or at least admission of anomaly, which helped
insure the survival of the system. Hutton, and many Huttonians,
had thought that the discovery of granite veins cutting across the
country rock would provide a refutation of Werner's notion of
primitive, aqueously precipitated granite. On the contrary, Werner,
in 1791, acknowledged the existence of granite veins and even
turned the phenomenon into evidence for his own theory of aqueous precipitations: if we accept the "humid formation" of granite

in its primeval form, he asked, may we not also take its existence in veins as evidence for the same process? (Werner 1791/1809:76–77).

Werner dealt in a similarly disconcerting fashion with the evidence of indurated and altered borders of "veins," offered by Huttonians as evidence that they were not aqueous but clearly resulted from the injections of material so hot that it had fused and altered the sides of the rock as it came in contact with it. "In places where this peculiarity occurs," wrote Werner, referring to the blurring of vein walls, mixing, and induration, "the rock has had a strong attraction for the substance of the vein introduced into the rent, and has become so intimately mixed with it" (Werner 1791/1809: 90). This is no more convincing than Hutton's doctrine of attractive and repulsive powers of matter (Gerstner 1968), but the subsequent reasoning is more interesting from the standpoint of the relation of theory to particulars. At any rate, continued Werner, the phenomenon is rare, and *"in general* the vein and the rock are separated from each other" (Werner 1791/1809:90). Here we have an instance of a perspectivism in which any particular, no matter how anomalous it seems, is denied the ability to overthrow the whole system. Werner and his students subscribed to what might be called the "doctrine of the preponderance of phenomena," whereby the general theory was shaped by the most common observations and not the most rare exceptions, which might be reserved for later explanation without the necessity of abandoning the theory. As a methodological precept, this was not far from Lyell's doctrine of uniformitarianism, whereby Lyell announced that since we have not yet exhausted the known or possible effects of existing causes, we have no license in our "infant science" "to recur to extraordinary agents" (Lyell 1830–1833:I, 6). In Wernerian terms, having not yet exhausted the "known or possible effects" of existing aqueous causes, the infant science of geognosy has no license to recur to extraordinary phenomena, like the "rare" mixing of vein borders with the deposited vein. Werner's natural-historical methods, and this particular approach to philosophizing, gave his system coherence and flexibility.

The works of Hutton and Werner reveal the intellectual milieus within which they developed. Hutton's geology was deistic and teleological, interested in the connections between geological phenomena and the physical and chemical behavior of the rocks and concerned with the relations of geology and natural philosophy. Werner's geognosy was a pragmatic, "rock-in-the-box" approach to the establishment of the stratigraphical succession; it continued the work of a natural history formally structured by a century of observation.

In Edinburgh and elsewhere in the first decade of the nineteenth century, a Huttonian theory (with notable deletions) was pitted against a neptunist theory built on Werner's geognosy. Initially successful, the neptunists were shown to be wrong in their basic mechanism. The subsequent decline of the neptunist theory has sometimes been mistaken, intentionally or accidentally, for the overthrow of Wernerian geognosy—with which it was not identical. The core of the latter was, as we have seen, not grand theory but observational method, not confidence in a body of knowledge but in a method by which knowledge might be achieved. It remained a strong tradition in European geology into the nineteenth century, and the changes brought about in Werner's original formulation of the geognostic succession of formations were accomplished by Werner's students, practicing geognosy as they had been taught by Werner.

The two schools continued to be viable opponents for many years, as we will see. The tension between approaches to geology through physical and chemical theory and through natural history remained, as did the associated tension over the proper aim for geological sciences. The debate over the worth of experimental evidence continued, and to this day the question of the uniformity or occasionalism of geological agencies, both in kind and intensity, remains open.

Many of the substantial questions on which the schools were opposed remained the basic questions of geology for the next hundred years and more: continental uplift versus fluctuations in the level of the sea; dislocation through igneous intrusion into the sediments or through gravitational adjustments around the core of the earth; the distribution and amount of the earth's internal heat; the actual character of the ocean and atmosphere in earlier ages of the history of the earth; the process by which rocks became hard.

Opposition both on methodical and substantive questions divided geologists into two generally opposed camps well into the nineteenth century. While it becomes more difficult in succeeding decades to associate these agreements point for point with Huttonian and Wernerian views, this is as it should be: geology was not merely chasing its tail but moving to substantial refinements in the particulars of various positions—while retaining both the flavor and the fundamental oppositions of the early division of the science.

Nowhere in the history of geology is there a more concentrated and intense repetition of these themes than in the history of the study of mountain ranges. In subsequent chapters, when the history of the debate over the origin of mountains is explored, the resurgence of basic "Huttonian" and "Wernerian" positions will not always be explicitly discussed, but as will be seen, they are never very far in the background.

CHAPTER THREE

Elie de Beaumont and the First Global Tectonics

Between 1810 and 1830 support for the theory of aqueous origins, especially the idea that the primitive formations were precipitates from solution, continuously declined as increasing evidence suggested the greater adequacy of some form of the igneous theory. After 1820, basalt having been widely accepted as an igneous rock, a new willingness emerged among Wernerians to accept the importance of the earth's internal heat—volcanism, plutonic injection, and other igneous causes—in the formation and consolidation of the outer crust of the earth. Cordier's experimental work on basalt, George Poulett Scrope's (1797–1876) careful investigations of volcanic phenomena, Buch's observations in the Canary Islands, Scandinavia, Italy, and Germany, overwhelmed the Wernerian aqueous theory completely.[1]

The systematic study of stratigraphy, which Werner had done so much to develop, advanced in this same period. Georges Cuvier and Alexandre Brongniart's *Essai sur la minéralogique des environs de Paris, avec une carte géognostique et des coupes de terrain* (1811, 1822); William Smith's maps and studies of the strata of England, Wales, and Scotland; and Cuvier's *Recherches sur les ossements fossiles* (1811) and *Discours sur les révolutions de la surface du globe* (1811, 1817) were particularly notable because they not only revealed errors in the Wernerian succession of formations but established correlations of

---

1. However, Benjamin Silliman was still teaching the system in the United States as late as 1829 because in spite of its admitted flaws no better general introduction to the science had emerged in the intervening years (Ospovat 1960).

formations by their fossil content (paleontological stratigraphy) as a necessary adjunct to correlation by mineralogical and petrologic character (lithologic stratigraphy). Werner's transitional strata, introduced to update his original succession from the primitive to the sedimentary formations, proved the undoing of the fixed succession—attempts at interregional correlation of this group revealed them to lie between different horizons in different areas. In Scotland, the "transitional" are pre-Devonian, in Cornwall and Saxony pre-Permian, and in Switzerland, pre-Pliocene (Bailey 1967:77).

The extremely rapid development of stratigraphy and paleontology throughout Europe and Great Britain in this period provided a body of new material for the interpretation of the history of the earth; these decades have often been referred to as the "heroic age" of geology, in which basic units of the stratigraphical succession began to be established in a form we now recognize, especially in the works of English geologists.

Among the many famous textbooks that appeared in the 1830s organizing and developing this material, the most famous was certainly Charles Lyell's *Principles of Geology*, which appeared in three volumes between 1830 and 1833. Lyell's work marked the renewal of a general theoretical debate on the system of the world as exemplified by the phenomena of geology. *Principles of Geology* included an eminently readable and informed summary of the observations of the preceding decades and much original work on the stratigraphy of the Tertiary period. But it also contained more controversial material—particularly Lyell's extreme insistence on the secular uniformity of geological action and a new attempt to give special meaning to the term *geology* and thus resolve the terminological confusion (and methodological debate), which extended well back into the previous century, concerning the proper position of geology relative to other sciences.

Lyell's theory of geological process and his definition of the proper aim of geology were, of course, interwoven. For Lyell, geology was to stand in relation to the physical sciences as history does to the moral sciences. Properly understood and practiced, geology was to be the application of the sciences of chemistry, natural philosophy, mineralogy, zoology, comparative anatomy, and botany to the elucidation of the history of the earth. The extremes of making geology a subdivision or "subordinate department" of mineralogy or of cosmology were to be avoided. The former was too constricted in scope and had been Werner's "error"; the latter, the mixing of geology with cosmology, made the subject too broad, and was "the most common and serious source of confusion" (Lyell 1830–1833:I, 2–4). Lyell wished to avoid both myopic overconcentration on mere taxonomy and fantastic speculation and gratuitous hypotheses by finding for geology a middle ground—more than a study of rocks and minerals, less than a theory of the origin of the globe (or the universe).

As an applied historical science, built on a foundation of many independent sciences, geology would have to make its inferential structure conform to the logical development of the sciences from which it drew. Implicit in physics and chemistry and explicit in botany, comparative anatomy, zoology, and mineralogy was the argument that the past is known only with reference to the present. Fossils were interpreted as petrified remains of former organic life by comparing them with living mollusca and the bones of living animals: paleontology and comparative anatomy worked from the present into the past. Similarly, the basalt controversy had been solved by the study of active volcanoes and the composition of their lavas and ejecta (Lyell 1830–1833:I, 2–4).

Most of the folly attending speculative reconstructions of the history of the earth was occasioned, Lyell suggested, by an ignorance of the way the earth operates in the present and an ignorance of the content and explanatory power of the various sciences. The major thrust of *Principles of Geology* was a specification of the nature of the existing causes of geological change, their known and possible effects, their development, and their place in a theory of the earth that explained without recourse to "extraordinary agents," that is, to any process not now visibly operating. Only when the known and possible effects of existing causes were exhausted would it be permissible to "recur to extraordinary agents" in order to complete the picture (Lyell 1830–1833:I, 6). Since his was an "actualistic" theory, Lyell wished to set aside all theories of the development of the earth (including Werner's) that argued for nonuniform processes of change or wide variations in the intensity of action of known causes.

Lyell saw the face of the earth shaped by the ceaseless interplay of aqueous and igneous phenomena. The action of water from above constantly worked to level the earth, and the action of heat from beneath simultaneously worked to restore the inequality of surface elevation, "depressing one portion, and forcing out another of the earth's envelope" (Lyell 1830–1833:I, 167). The action of water was expressed in the erosion cycle, the work of rivers and springs, tides and currents, wearing down the land and carrying it away into the sea to be deposited anew. The character and location of deposits were controlled by the distribution of land and sea, which in turn, produced regional variations in climate (Lyell 1830–1833:I, 104–105). New distributions of land and sea and the resulting climatic variations were further determined by the uplift and subsidence of the crust brought about by earthquakes and volcanic action, which were the joint expression of the same subterranean process—the release of the vast store of the earth's internal heat. Lyell assumed that at some distance below the surface the earth was molten under great pressure and thus exerted a hydrostatic pressure on the crust from beneath "capable of moving enormous masses of land" (Lyell 1830–1833:I, 463). Lyell rejected Hutton's idea of the "waste of successive continents followed by the creation of others by paroxysmal convulsions." Rather, in his view, the repair was as constant as the decay. Volcanoes and other thermal phenomena were, as in Hutton's work, "safety valves" releasing excess internal heat, without which the subterranean buildup of heat would bring on perpetual convulsions of the crust (Lyell 1830–1833:I, 470).

This theory was quite close to Hutton's, but Lyell developed it in a new direction. In the Huttonian theory there was no paleontology, no study of the history of life. For Lyell, life history was the key to the historical elaboration of much of the earth's geological development. The action of the various geological agencies determined the nature and distribution of former life forms by altering the areas of land and sea and thus forcing variation in climate. Therefore, the study of past life forms was essential to documenting these former distributions of land and sea and their climates. Beginning from the way animals lived in the present—on land or in the sea, in warm or cold regions—one traced these forms back to similar animals in earlier epochs and used these morphological similarities as a basis for geological inferences about geomorphology and climate. Lyell's researches enhanced his conviction that geological activity, whether constructive or destructive, proceeded regularly and in small increments through immense stretches of time with little variation in intensity and almost no variation in kind; for him the earth was steady, stable, and self-perpetuating.

The controversy following the publication of Lyell's book is one of the most chronicled episodes in the history of geology.[2] Most geologists in the 1820s believed that the earth had been subject

2. See, for example, Gillispie 1959; Cannon 1960; Hooykaas 1963, 1970.

periodically to episodes of great violence in which normal igneous and aqueous agencies were swollen to catastrophic proportions: great volcanic upheavals, huge floods and deluges. However, by that time, the invocation of these cataclysmic irruptions had little to do with biblical conformity, especially outside Great Britain. Cuvier employed them as an explanation for the periodic extinctions of whole classes of life forms revealed in his paleontological work, and many others followed suit. The idea of a great flood, or many such floods, was particularly useful in explaining the distribution of enormous amounts of loose sand, gravel, rock debris, and erratic boulders over much of the surface of Great Britain and Europe. The erosion cycle was not useful as an explanation, for these materials often consisted of rocks with no petrological relationship to the country rock over which they were strewn—nor did they appear to be the result of local upheaval. Certain geomorphic phenomena —deep ravines, gaps in mountains, isolated buttes—also seemed to be best explained by intense forces, especially flooding, acting over a short time (Gillispie 1959:110).

Catastrophic theories came into being in the first place as answers to otherwise inexplicable paleontological and geological evidence. The development of paleontology weakened the argument for catastrophic extinctions in that the phenomenon of extinction was seen to be pervasive through geologic time and not isolated; but the explanation of the "diluvial gravel," as the erratically distributed surface materials were called, was more difficult. Buch was able to show that many of the huge erratic boulders on the North German plain had originated in Scandinavia, and he explained their transport by means of a catastrophic flood. Against this theory, Lyell and other British geologists proposed a "drift theory" in which the materials were transported by drifting icebergs during periods when the distribution of land, sea, and climate allowed such a thing to happen (Zittel 1901:230–231). Eventually, toward the end of the century, these phenomena were explained by the continental glaciation theory of Agassiz and Ramsay. However, in the 1820s and 1830s, field evidence seemed to require geological action of an intensity far beyond Lyell's definition of existing causes and seemed to throw doubt on the adequacy of his theory of nearly uniform change.

In Great Britain the problem was further complicated by the continued exploitation by physico-theologians of the "diluvial gravel" to substantiate the adequacy of the biblical account of the earth's history, particularly the great Mosaic or Noachian deluge. Although the importance of this movement on geological thought in the 1830s has probably been exaggerated, distinguished geologists were still attracted to the banner (Page 1969:257). William Buckland's *Reliquae Diluvianae* (1823) and Charles Daubeny's *A Description of Active and Extinct Volcanoes* (1826) used the idea of catastrophically powerful floods and volcanic outbreaks (respectively) as the means by which the Creator had employed His creation in the exercise of His will, and they were able to put their geological researches to the services of revealed religion (Gillispie 1959:ch. 4).

The debate over the rate of change in geology spurred by Lyell's theory occupied almost all the major figures in British geology—Buckland, Daubeny, Adam Sedgwick, Roderick Murchison, William Whewell, Henry De la Beche, William Conybeare, and of course, Lyell. In the debate, conducted within the confines of the Geological Society of London, the physico-theological aspects quickly ceased to be of much importance—with geologists quickly backing away from the assertion that any portion of the diluvial gravel could be correlated with the specific event of the biblical flood. But even the more theologically oriented of Lyell's opponents remained unconvinced, on geological grounds alone, of the extreme uniformity of nature he had postulated.

This continuing division has come to be known as the Uniformitarian-Catastrophist Debate, but the terms *uniformitarian* and *catastrophist,* made popular by William Whewell in the 1830s, are even more deceptive than vulcanist, neptunist, and the other factional labels applied in earlier decades because they hide a wide variety of opinion on the nature, intensity, and relative uniformity or periodicity of geological phenomena. To use the terms is to arrange mid-nineteenth-century geological theory into two groups —those who supported Lyell and those who opposed him—thus giving the impression that Lyell's work was of paramount importance throughout the period; it was not. A geologist like Karl von Hoff (1771–1837) might espouse a theory of uniform change not as a minion of Lyell or out of reverence for a philosophic principle but because the evidence he saw seemed to argue that way (Hooykaas 1963:4–6). Even so, *uniformitarianism,* however arrived at, is much more homogeneous and therefore less misleading than the term *catastrophist.* The theories of the many proponents of historical catastrophic events varied across a wide spectrum of assumed violence and had an equally varied content. Some geologists held to the idea of "ancient causes"—geological phenomena of great violence in the past history of the earth that for a number of reasons could no longer occur, but which had left their marks on the crust. Others argued that regular phenomena such as earthquakes and volcanic eruptions sometimes acted with catastrophic intensity. Still others argued that phenomena different in kind from the ordinary course of events sometimes appeared—the upthrust of mountain ranges, visitation by comets, shifts in the earth's axis resulting in floods. None of these opinions was strictly incompatible with the ideas of the erosion cycle and the alternations of land and sea proposed by Lyell, and it is therefore important to understand that catastrophism was not per se incompatible with a rational and uniform account of the progress of the earth through time. What the so-called catastrophist theories expressed in common was a disinclination to accept Lyell's invocation of the current state of the life of the earth as an adequate explanation for everything that might have happened in its past history. Particularly at issue was Lyell's assertion that the earth showed no hint of progressive development (Hooykaas 1970).

This was the great dividing question—whether geology revealed a cumulative development or a mere repetition of events through time culminating in a secular, perhaps eternal status quo (Cannon 1960:39). Lyell's presidential addresses to the Geological Society of London in 1850 and 1851, and William Hopkins's address in 1852, affirmed this basic ground of difference at mid-century (Cannon 1960:53).[3] In spite of wide agreement on specific dynamical phenomena, especially the action of ice and the importance of erosion and sedimentation, no fundamental reconciliation could be achieved.

Most of Lyell's opponents saw some progressive tendency in the history of the earth, and particularly prominent was the idea that the earth's geological history reflected progressive cooling from an original incandescent state. Lyell had attempted to remove this cosmogonic deduction from the list of reputable subjects that a geologist might treat, but many geologists refused to acquiesce.

In the group that declined to accept Lyell's theory and its limitations were influential geologists who devoted much or all of their professional careers to the exploration and explanation of mountain ranges, and they noted that the great connected mountain systems of the earth were a considerable embarrassment to uniformitarian geology. Lyell devoted little space in his *Principles of*

3. Cannon's ideas have been developed further by Martin J. S. Rudwick (1971: 209–227). I did not see Rudwick's work until I had worked out my own ideas on the subject and am pleased to corroborate his suggestion independently.

Geology to the genesis of mountains and implied that even such ranges as the Alps and the Andes had risen to towering heights by incremental stages of a few inches or a few feet at a time, lifted by earthquakes and volcanism (Cannon 1960:42; Lyell 1830–1833:I, 137–141). Even this was shaky ground for uniformitarians, since it forced acknowledgment of incremental and cumulative action rather than random repetition; Lyell preferred to avoid the topic and never developed a satisfactory theory of mountain ranges.

This lacuna was a great barrier to the acceptance of Lyell's geology in Europe, in spite of prominent French geologists like Ami Boué (1794–1881) and Constant Prévost (1787–1856) who welcomed his avoidance of catastrophism. The major question in European geology throughout the nineteenth century was precisely the origin of the great mountain ranges—the Alps, the Jura, the Apennines, the Carpathians, the Balkans—all of which together dominated the continent; no theory unable to explain them had a chance of success. Here, a historical overemphasis on the philosophical and theological debate in Great Britain has served to hide the general trend of geology; while British geologists debated change, the study of mountain ranges was advancing throughout Europe and the New World, and in the forefront of the first wave were the students of Abraham Gottlob Werner. After them came a generation of students in the same tradition, some of whom traced their inspiration back directly to the professor at Freiberg. In their work we see the survival of many elements suggested in the work of Werner and the Wernerian school, most particularly in the persistent attempt to develop laws of mountain structure on a foundation of geometry.

The study of the geometrical arrangement of the earth's surface features was given new prominence as an important field of geological research in the work of the enormously influential French geologist Léonce Elie de Beaumont (1798–1874), Lyell's exact contemporary. Elie de Beaumont's ideas on the causes of mountain formation and the way in which mountain ranges should be studied dominated continental geology at mid-century and were the principal reason why Lyell's complete theory never gained a firm foothold in Europe during his lifetime. His education and career reveal the way in which European geology proceeded directly from the foundations laid by Werner, Saussure, and other European geologists of the early period toward a culmination in the work of Eduard Suess, Marcel Bertrand, and other famous Alpine geologists at the end of the century, without being significantly influenced by British ideas, either Huttonian or Lyellian.

As a young student Elie de Beaumont was a gifted mathematician. He graduated with honors in mathematics from the Collège Henri IV in 1817, and in 1819 he graduated at the head of his class from the Ecole Polytechnique. His choice of the Corps des Mines was unusual for a student with such gifts—not a single Polytechnique student had entered that service in the three preceding years. He immediately undertook geological fieldwork in preparation for his career, applying himself to researches in the stratigraphy of the Vosges region of France and in the portion of the Jura Mountains that lies in Switzerland. He was chosen to accompany the French delegation to England to learn cartographic technique preparatory to the commencement of the long-delayed "Carte géologique de France"; upon his return he was set in charge of the mapping of the eastern (and most mountainous) part of France.

Elie de Beaumont's rise to scientific prominence was rapid, as was his professional advancement in the Corps des Mines. He was promoted to Ingénieur des Mines in 1824, and in 1827 he began to lecture at the Ecole des Mines. By 1828 he was a member of the Société d'Histoire Naturelle de Paris, an honorary member of the Société Helvétique d'Histoire Naturelle, and a correspondent of the Royal Prussian Academy and other groups. In 1830 he helped to found the Société Géologique de France, and in 1832 he took up teaching duties at the Collège de France, where he remained "Professeur titulaire de l'histoire naturelle des corps inorganiques" for the rest of his life (Birembaut, 1970:350). He was also named director of the Carte Géologique, making him the most powerful French geologist of his time.

In preparing his first publication, *Coup d'oeil sur les mines* (1824), a survey of mineral resources, Elie de Beaumont was able to familiarize himself with the mountain ranges of the world while assessing their exploitable contents and discussing the techniques employed to recover them. Mining centers were considered by continent and region, located by latitude and longitude, and grouped by occurrence in primary, secondary, and alluvial terrains. In addition to the more familiar locales of England, France, Germany, and Spain, the survey included catalogues of mines and resources in Brazil, Tibet, China, Siberia, the Congo, and other relatively unknown and exotic locations. At this early stage in his career he was able to learn something of ranges to which he and other European geologists had had little access—the Altai, Alleghenies, Urals, Andes, Mexican Sierras—as well as the Harz, Erzgebirge, Apennines, Pyrenees, Jura, and the Alps.

In a brief appendix to the survey, Elie de Beaumont paid homage to those great geologists of the preceding generation who had also been miners—Werner, Haüy, and Dolomieu. Of great note in Werner's work was his general discussion of the trend of mineral veins and their geometrical arrangement: "the laws of their parallelism, of their intersections" (Elie de Beaumont 1824:142). These "laws" appeared in Werner's *Neue Theorie von der Entstehung der Gange* (1791/1809), which, in addition to a full version of his (incorrect) theory that veins were fissures filled from above, also discussed the means of tracing mineral veins through complex formations. Werner had written, "It is natural to suppose, that when a mountain is rent at several different times, it has been done at each of these by the same force acting in the same direction; so that all the veins which have been formed at the same time ought to be parallel or nearly so" (Werner 1791/1809:97). Whereas the English translator found that "Werner had a manner of expressing himself peculiarly his own, and in many of the most valuable parts of his subject frequently makes use of a tautology which appears unnecessary" (Werner 1791/1809:x), Elie de Beaumont, the young mathematician, saw rather an attempt at rigorous ordering of propositions employing a geometrical style of reasoning from axioms and postulates.

This is the more evident because Elie de Beaumont also chose to celebrate the work of Haüy. René Just Haüy (1743–1822) had taught both physics and mineralogy at the Ecole des Mines beginning in 1795, and in 1809 assumed the new chair of mineralogy at the Sorbonne (Hooykaas 1972:178). His achievement in mineralogy was the application of mathematics. He developed what was later called the crystallographic law of rational intercepts. The law states that when referred to three intersecting axes, all faces of a crystal can be described by numerical indexes, which are integers and usually small numbers. While the law was purely empirical, "the assumption of its universal applicability was fundamental in the mathematical approach to crystallography" (Burke 1969:78–79). Moreover, Haüy's application of mathematical techniques had ordered a field of study marked in the preceding century by numerous failed attempts at a general system (Metzger 1918). In mineralogy, it allowed another avenue to identification and classification of minerals, less cumbersome than either external characters or chemical analysis. Haüy's great *Traité de crystallographie*, a development of work begun in 1784, was published in 1822, and "the genius that had created this valuable scientific theory from so small a number of empirical facts was appreciated fully" (Burke 1969:106).

The third great miner-geologist honored by Elie de Beaumont was Guy S. Tancrède de Dolomieu (1750–1801), a student of mountain structure (and a neptunist) who had written on the composition of crystalline rocks, as Elie de Beaumont was later to do himself. A professor at the Ecole des Mines after 1796, he was later professor of mineralogy at the Musée d'Histoire Naturelle in Paris. He espoused a theory of the catastrophic uplift of mountain ranges as a necessary deduction from the evidence of field geology. Dolomieu was not a catastrophist on principle, but it seemed to him that the slow action of the ocean or any other agency could not possibly be responsible for the magnitude of the disruptions evident in mountain ranges. He wrote to Saussure that he had no attachment to the hypothesis of great catastrophes, but whatever hypothesis was put forward to explain the structure of mountains should offer "a cause sufficiently active for producing the required effects" (Hooykaas 1970:15–17).

Taken together, these three geologists provided Elie de Beaumont with a foundation for his entire approach in subsequent works: from Haüy, the inspiration for the application of simple mathematical techniques to a complex (and otherwise heterogeneous) set of phenomena; from Dolomieu, the conviction that the ordinary intensity of action of geological causes could not explain the genesis of mountain ranges; finally, from Werner, the notion that the parallelism and intersection of geological structures was an important tool in unraveling their history and genesis. In this joint celebration of earlier masters can be seen the beginnings of Elie de Beaumont's own unusual blending of geology and geometry, and the sense that application of mathematical techniques to geological problems was a fruitful area of research, likely to produce further great results.

The application of geometry to geology had a tradition extending back to the time of Agricola and even earlier. "Subterranean geometry"—the gauging of the strike and dip of sedimentary beds, the angles of intersection of mineral veins, the thicknesses of the sedimentary beds—had long been employed to generate empirical rules useful in locating and measuring the extent of mineral deposits. In addition, the study was an economic and practical necessity in hard-rock mining and shaftworks, where the digging was laborious and dangerous.

However, in the seventeenth century, more ambitious and speculative applications of geometry had been attempted with respect to large-scale features of the crust, as in the work of Steno and in

the work of the Jesuit philosopher Athanasius Kircher (1601–
1680). It was a common contemporary conceit that the mountains
were the skeleton of the earth's body, and in *Mundus Subterraneus*
(1664) Kircher made an attempt to show that they intersected in a
regular pattern (Adams 1938:433; Nicolson 1959:168–172). This
sort of crude intuition was eclipsed by more painstaking observations in the next century, particularly those of Alexander von
Humboldt (1769–1859).

Humboldt had been a student of Werner and was a great geological explorer, having traveled to the Americas and Asia in addition to his extensive forays in Europe. In the 1790s he developed
the hypothesis of the parallelism, or "loxodromism," of the strata
of the mountains of Central Europe. His idea was that regardless of the geographical direction in which a group of mountains
tended, the strike and dip of the strata (their geographical alignment and their angle with the horizon, respectively) were a function
of their age (Baumgartel 1969:30). This was an early realization
that tectogenesis (the origin of structure) and orogenesis (the origin
of mountains) might be separate phenomena. One of his purposes
in traveling to the New World was his hope to confirm this law of
loxodromism. When his researches failed to provide the desired
corroboration, he gradually reformulated the insight away from
the strike and dip of the component beds toward the general trend
of surface features, especially mountain ranges; he thought he saw
in mountain chains a preferential direction of the long axes from
northwest to southeast. Humboldt was convinced that this coincidence in the trend of mountains was in some way an expression of
fundamental laws of development of the earth's crust (Baumgartel
1969:30). Elie de Beaumont had studied Humboldt's observations
and relied on them heavily in works published in 1824 and 1828,
particularly on Humboldt's descriptions of South America (Elie de
Beaumont 1824:68; 1828).

During the time of Elie de Beaumont's schooling and first fieldwork there had been more general confirmation of some meaning
in the direction of mountain chains. In 1816, the editors of the
Bibliothèque universelle had published extracts from de la Métherie's
geological lectures at the Collège de France. De la Métherie had
speculated that the great Alpine peaks were crystals, a theory that
Adams later argued was "surely one of the most fantastic that ever
entered the mind of man" (Adams 1938:388). The publication of
the extracts was accompanied by some rather strong criticisms of
the ideas expressed, especially the idea that mountains could be

individuals in any sense. It seemed, wrote the editors, to have escaped the attention of most geologists that with very few exceptions
mountains occurred as chains and not in isolation. Further, the
existence of a chain of mountains supposed an action similar in
effect along an extended line—the existence of chains as longitudinal projections in the geometrical form of prisms rather than isolated cones was certainly worthy of more serious attention than it
had heretofore received (Anon. 1816:30).

In 1819, J. F. d'Aubuisson des Voisins published his synoptic
text *Traité de géognosie*. In it he proposed a law of the direction of
mountain chains: "In general, mountain chains extend in the direction of the greatest dimension of the islands, peninsulas, or continents where they occur" (d'Aubuisson des Voisins 1819:101). His
aim was to correct the idea (that arose again later in the nineteenth
century) that they followed parallels and meridians in a gridlike pattern, but he nevertheless imposed a determinate structural pattern
on mountain genesis. He had also been a prominent, though independent, supporter of Werner, and in the same work he underlined the isolation of the British tradition of Hutton, Playfair, and
Hall from the mainstream of European geology, in which France
and Germany shared common Wernerian leanings. Developments
in France (influenced by his own position as chief engineer of the
Corps des Mines) were predominantly the modification and extension of Werner's teaching and included many changes that he had
himself instituted (d'Aubuisson des Voisins 1819:xvi). Dolomieu
and Haüy were part of this extension of Wernerian geology, as was
Brochant de Villiers. De Villiers, incidentally, was professor of
geology at the Ecole des Mines, and Elie de Beaumont's patron. It
was through his influence that Elie de Beaumont made the trip to
England in 1824 and was later named de Villiers's successor as
inspector general of mines in 1835 (Zittel 1901:299).

Thus, the tradition of Wernerian geology in France was propounded by all of Elie de Beaumont's teachers and patrons, his
predecessors in a series of controlling offices within French geology, who certainly shaped his views on the mainstream of geological thought. Even so, their teaching did not represent the highest
development in the study of mountains achieved in Europe to that
time. D'Aubuisson des Voisins's views were already too schematic to
merit much attention in Switzerland where, as one might expect,
the study of mountains and geology were nearly synonymous. In
the eighteenth century a number of researchers had found a cross-
sectional similarity in major mountain ranges in Europe—a central

zone of crystalline rock, flanked bilaterally by deformed sequences of sediments; their researches were prominent in the development of the Wernerian stratigraphic succession. Among the most celebrated and influential in applying this insight to Swiss mountains were Horace Bénédict de Saussure (1740–1799), Johann Gottfried Ebel (1764–1830) and Hans Konrad Escher (1767–1823).

Saussure, whose Alpine studies were prominently featured in the explication of both the Huttonian and Wernerian systems, made numerous independent traverses across the Alps, and his work marks the real beginnings of Alpine geology. A long-time supporter of Werner, Saussure became convinced by his own observations that the neptunist theory could not be entirely true. In the neighborhood of Valorsine, Saussure observed a conglomerate (a rock containing pebbles and fragments of preexisting rock in a solid matrix) that the British called puddingstone and the French, delightfully enough, *poudingue*. Saussure found a vertical section of this conglomerate that contained a large boulder. In the Wernerian scheme, the occurrence of such a formation would have been explained as a fissure filled from above, thus accounting for the unavoidable evidence of rock fragments in a matrix. Saussure, who had vacillated between neptunist and plutonist ideas while confronted by the complexities of Alpine structure, was finally unable to see how such a large rock could have remained suspended in place long enough for the infilling fluid medium to harden around it, and he took the exposure as evidence against the Wernerian theory. The bed must originally have been a flat stratum, he reasoned, and later tilted into position in a subsequent dislocation. His work was never systematized into a theory of the Alps but was published and widely circulated as *Voyages dans les Alpes*, from which Hutton took massive sections for the second volume of his *Theory of the Earth*.

Hans Konrad Escher was important for the technical detail he pursued in his work and replicated on paper. The sections he made in preparation for the construction of the Linth Canal were of such importance that his son Arnold, also a premier Alpine geologist, carried the name Escher von der Linth in commemoration of his father's pioneering work. The senior Escher also prepared a survey of the Alps in detailed sections from Zurich to the St. Gotherd Pass. Escher's sections in combination with the work of Saussure were the foundation of the general account of the Alpine system by Johann Gottfried Ebel, published in 1808, *On the Structure of the Earth in the Alpine Mountain System*.

Ebel's work began with the idea of a threefold structure—a central crystalline zone flanked by sediments to the north and south—and he developed the stratigraphic sequences carefully and in detail. He also provided a map comparing the Alps with the other mountains of Europe and worked on the structure of the Jura range. The Jura Mountains were also investigated by Escher, who presented a paper before the Société Helvétique d'Histoire Naturelle in 1820 on the structure of the chain as a series of parallel ranges, the ensemble of which were parallel to the Alps (Escher 1820:75). Ebel strove for a theoretical interpretation even broader than Escher's and tried to work the Jura into a geometrical system that accorded with the laws of loxodromism (of preferential geographic orientation) proposed by Humboldt (Zittel 1901:92).

But among all early nineteenth-century theories of mountain structure, the most influential by far were those of the German geologist Leopold von Buch (1774–1853). Buch, like Humboldt, was a student of Werner and to the end of his life a great traveler and avid field geologist. He figured prominently in the debate between the neptunists and the Huttonians over the origin of basalt, and he is supposed to have converted to the plutonist cause after an inspection of the volcanoes and lava flows of the Auvergne region of France. He did not, but after his trip to the Canary Islands in 1815, where he studied the relief of lava pressure in the interconnected cone system of those islands, he was more and more attracted to the idea of great subterranean lava reservoirs that would provide the expansive power for the uplift of mountains. His travels in the southern Alps (the High Dolomites) suggested the typical aftermath of such an upheaval (Nieuwenkamp 1970: 552). To the modern observer the High Dolomites, irregular stark pillars surrounded by talus, suggest erosion and frost-shattering. Buch, however, saw in the Dolomite peaks the evidence of the catastrophic upthrust of the pillars, with the talus as the debris of that forceful elevation. It is quite consistent with his own scheme of things, and only curious in retrospect, that Buch, in spite of his great reputation as a geologist, never accepted the idea that erosion significantly affected surface relief (Nieuwenkamp 1970:557). Buch might have become convinced of the importance of plutonic injection and upheaval, but his refusal to accept erosion makes it impossible to consider him a convert to Huttonianism.

Buch's theory of the elevation of mountains—the theory of "craters of elevation"—became one of the two most important orogenic theories of the nineteenth century. He attributed the existence of mountains to the upwelling of molten matter from beneath. Volcanoes were isolated uplifts in which a rising column of molten rock radially elevated a portion of the crust, first raising a blister, and then breaking through its apex and overtopping it with flows and explosive gouts of lava. In the case of mountain chains, the molten material welled up along a line that determined the axis of the resulting chain. In both cases, the strata were split apart and deformed from the center outward, either radially around centers of elevation (*Erhebungscentra*) or bilaterally along axes of uplift. In this theory the crystalline rocks at the core or apex of many European ranges, originally interpreted as the primitive precipitations on the core of the earth, were seen as later, catastrophic upthrusts of plutonic masses. The idea found great favor among Alpine geologists and was also applied in the United States to the Green and White Mountains of New England (Vose 1866:17–18).

In 1824 Buch published the results of his study of the mountains of Germany based on this line of reasoning: *Ueber den Geognostischen Systeme von Deutschland*. He concluded that there had been four separate catastrophic outbreaks, resulting in four separate mountain systems, each one with its own trend or direction and each of a different age. This correlation of the age and the strike of a chain, with the latter as a diagnostic feature of the former, became one of the principal elements in the mountain theory of Elie de Beaumont (Bailey 1935:11, 14).[4] Charles Sainte-Claire Deville, a student of Elie de Beaumont and his deputy at the Collège de France, saw Elie de Beaumont and Buch as developers, in slightly different directions, of "the great thought of Werner in which consists his most abiding fame: the law of parallelism of directions in geological phenomena of the same age" (Sainte-Claire Deville 1878: 354).

Up to the year 1829, Elie de Beaumont had produced no work of a theoretical character. He had concentrated on his survey of mines, the progress of the Carte Géologique, his stratigraphical paleontology of the Vosges region, and his teaching duties at the Ecole des Mines and the Collège de France. In 1829, while Lyell was preparing *Principles of Geology* for publication, Elie de Beaumont read his first publication on the theory of the earth to the Acad-

---

4. Bailey, a director of the Geological Survey of Great Britain, was a supporter of the nappe theory in the Alps, of continental drift, and of the nappe theory in the northwest Highlands of Scotland. His informal history of the Geological Survey of Great Britain (1952) is very useful on the history of geological theory, and his biography of Hutton (1967) is also valuable.

émie des Sciences. His audacious thesis involved a correlation of the paleontological theories of Georges Cuvier with Buch's episodes of catastrophic mountain uplift (Elie de Beaumont 1831).

Cuvier had offered the theory in his *Discours sur les révolutions de la surface du globe* that there had been six successive catastrophes in the history of the earth marked by widespread extinction of life forms. Buch had postulated successive and destructive upheavals of mountain ranges, and Elie de Beaumont wanted to identify the massive extinctions with the geological upheavals. Must not, he asked, the "frightful convulsions which must have accompanied the upthrust of masses so great, and of an aspect so contorted as those of high mountains" have been accompanied by other revolutions? Elie de Beaumont was not attached to the particulars of their theories. After all, there were six catastrophic episodes in Cuvier and only four in the work of Buch; nevertheless, their work together, in principle, established for him the two great generalizations of geology; epochal disturbance of the course of organic life, and simultaneously of the secular processes of geology. From these generalizations he essayed a theory of geological history opposed to that contemporaneously developed by Lyell.

Lyell, of course, was not yet well known and did not figure in Elie de Beaumont's thought as an antagonist, nor did Hutton. The historical prelude to Lyell's theory ignored the survival of any antagonistic schools of geological thought. However, he did implicitly dissociate himself from that part of Werner's views concerning the origin of sediments, turning instead to the work of Niels Steensen and Horace Bénédict de Saussure. Steensen was the late-seventeenth-century Danish physician and geologist who had published a geological and paleontological essay on the area surrounding Tuscany. He had argued that strata were deposited horizontally on the ocean floor and that their broken dislocated appearance was the result of subsequent activity; this was also the later conclusion of Saussure.

Elie de Beaumont's reading and his own experience led him to extend Saussure's observation to all mountain ranges that were composed of folded and tilted strata like the Alps and the Jura. These same mountains, he had observed, had at their feet broad horizontal plains of undisturbed strata. This contiguous occurrence of contorted and undisturbed strata could be explained by assuming that the upthrust strata had subsequently formed the shore, or part of the shore, of a lake or ocean in which the flat-lying strata had been peaceably deposited. It was clear from fossil evidence that the flat strata were of a later age than the contorted mountain strata and that the latter could not possibly have been upthrust through them. Thus the fixing of the age of the most superficial formation on the flanks of the mountain chain, and of the deepest horizontal formation contiguous to it, marked in the history of the earth an episode of mountain building that had interrupted the normal succession of depositions. "In each system of chains, the series of sedimentary rocks is divided into two distinct classes, and the point of separation of the two classes, variable from one system to another, is the circumstance which best characterizes each particular system" (Elie de Beaumont 1831:242). Elie de Beaumont claimed no originality in seeing in such discordances an index to the dating of formations—the absence of parallelism as a signature of the geological time of their occurrence had already "become common."

Indeed, it had. Hutton in the previous century had taken such occurrences as evidence of the erosion cycle. The classic example for Hutton was the Siccar Point unconformity in Berwickshire, in which the formation now known as Upper Old Red Sandstone lies unconformably above the steeply tilted Silurian slaty mudstone. For Hutton, this was evidence that a period of deformation (the tilt) and erosion (the absence of intervening beds) had occurred before the superior formation could have been laid down. But in the Huttonian theory of the earth there was no separate account of mountain building as such. Uplift occurred, with dislocation, during continental elevation, and the height of mountains above surrounding country was not a phenomenon of *structure* but of *relief*. Following the uplift of a continental mass, the secular work of erosion carved out valleys and wore the land down. What appear in the present as mountains were for Hutton those parts of a continental mass that testify to the original scale of the uplift, and are those parts of it not yet worn down to the level of the sea. Mountain ranges are landforms but not special structures requiring any particular explanation. Deformations of strata, observed in mountain ranges, had nothing to do with any process of mountain building but were the result of great temperature and pressure acting on the strata close to the source of the earth's internal heat, in the period before the continent was lifted from the floor of the sea. The disjunction of fauna and the interruption of the regular series of deposits meant only that the tilted and the horizontal strata belonged to two different epochs in the history of a continental platform.

Elie de Beaumont thought otherwise. "It follows," he wrote,

"from this difference, always clear and without passage, between the uptilted beds and those which are horizontal, that the elevation of the beds has not been effected in a continuous and progressive manner, but that it has been produced in a space of time comprised between the deposition of two consecutive rocks, and during which no regular series of beds was produced;—in a word, that it was sudden and of short duration" (Elie de Beaumont 1831:243). It should be noted here that he asserted only that in certain mountain ranges structural discordances occurred that were not part of this secular mechanism and did not deny the erosion cycle, to which he subscribed. A conviction that there were some catastrophic processes in nature was not confused with or assimilated into a doctrine that all change was necessarily catastrophic. Not all tilted beds were the result of general catastrophes—but those in mountain ranges of the folded type were.

Fold mountains could not be mere erosion products, he argued, for they were not local phenomena. Their conformable structure for great distances along the strike of a chain could not be explained by slow and irregular action alone. That the tilt of beds remained constant for long and even immense distances in such ranges had been known for ages by miners. Field observation, he continued, by Werner, Pallas, Saussure, Buch, and Humboldt had confirmed the existence and the regularity of such phenomena. Buch's theory of the mountain systems of Germany (1824) was the latest evidence in favor of an interpretation in which the tilt of beds was ascribed to the upheaval of mountains from flat-lying strata.

In adopting Buch's theory of elevation of chains along an axis, Elie de Beaumont concurred with him in the notion of successive episodes of upheaval, with the direction of the upheaval (the geographical direction of the strike) indicating the time at which a range had been created. But the attempt to correlate the geological ideas of Buch and the paleontological ideas of Cuvier led him to a much larger generalization: the number of such events of uplift was not unlimited, but it was at least equal to the number of distinct directions of mountain chains in the world, a number, he urged, "not incompatible with the discontinuities" observable (Elie de Beaumont 1831:245). That is, the dynamic history of the crust could be inferred from concentration on the mountain ranges of the world. All ranges of the same age tended in the same direction, and this correlation seemed to be confirmed by the presence of great angular unconformities at the base of each range, the fossil contents of which might establish the date of the episode.

In pursuing this thought, Elie de Beaumont left the mountains of Europe behind and founded his argument on an overview of the earth as a whole, controversially abandoning close analysis of specific structures for inferences derived from structural affinities of the highest generality—the geometrical arrangement of mountain ranges over the surface of the globe. If we look at a good globe, he wrote, we can see that the most prominent and recent chains of Europe form a vast system of parallel chains—a system that extends into geologically unknown regions. The idea of the contemporaneity of parallel chains suggested to him that the Eastern Alps were "part of a vast assemblage of mountains which spread around the Mediterranean and [are] prolonged across the continent of Asia" (Elie de Beaumont 1831:260). This assemblage included, to the west, the Atlas (North Africa) and the mountains of Spain, and to the east, the mountains of the Balkan peninsula. Farther to the east, the Hindu Kush and the Himalayas seemed, on the basis of parallelism, to be part of the same system, spread along a single arc.

The Alps are the greatest chain in Europe, and their division into the Eastern and Western Alps was based on a lack of stratigraphical continuity between two major sections of the chain as a whole. Elie de Beaumont's structural hypothesis explained this disjunction by pointing to the fact that the Western Alps were not parallel to the Eastern but were part of an entirely different set of parallel chains extending to the east along a different arc and passing through the coastal mountains of Morocco to vanish beneath the Atlantic, only to reappear in South America as the Littoral Cordillera of Brazil. A third arc connected the Pyrenees and Apennines with mountains as far east as Malabar in the Indian Ocean, and as far west as the Allegheny Mountains in the eastern part of the United States. Turning from the Atlantic to the Pacific, he pointed to the existence of a ring of high mountains around the Pacific extending from Chile to Burma. This latter feature of the earth's crust was the largest of all the arcs he had considered, and the still-smoking Andes convinced him of the relative newness of the ranges along the coast of South America. He mused that this most extensive feature of the earth's crust might have been responsible for the historical event recorded as the biblical deluge. If a mountain range as large as the Andes were suddenly thrust up, the resulting tidal wave around the world could easily account for such a catastrophe (Elie de Beaumont 1831:261).

Elie de Beaumont was convinced that this arrangement of mountains in distinct arcs was due to a simultaneous uplift along "a considerable portion of one of the earth's great circles." The mechanism of this uplift was not known, but certainly the cause had to be as regular as the phenomenon, even if there were "disturbances different from the ordinary march of the phenomenon which we now witness" (Elie de Beaumont 1831:262).

In the spate of opinions known together as catastrophism, Elie de Beaumont was an "actualistic catastrophist"—whatever raised these mountains had to be explained within the realm of "reasonable causes." Reasonable causes were those that obeyed the known laws of nature and, further, were capable of acting in the present as well as the past: the upthrust of mountains could not be something from an earlier, uniquely paroxsymal epoch of earth history. He could not imagine that "the mineral crust of the globe has lost the property of being successively ridged in various directions" (Elie de Beaumont 1831:261). But what, then, in the ongoing mechanism of the earth, could explain such phenomena?

He considered the possibility of volcanic action as a cause. Urged by Buch and others, it was the most obvious agency—the Andes, which were the largest and most recent range, contained numerous active volcanoes, and there was ample evidence of their power. But he rejected the idea. Even though (as Humboldt had shown) volcanoes were sometimes arranged in lines, following fractures parallel to mountain ranges, their radial form did not seem to be conducive to the uplift of long ridges "which follow a common direction through several degrees of longitude" (Elie de Beaumont 1831: 263). Contrary to Zittel's attribution, Elie de Beaumont did not in 1829 accept the "craters of elevation" theory of Buch (Zittel 1901: 264). For an explanation of the mountain uplift he turned to an idea that he ascribed to Humboldt: the slow cooling of the earth, "the slow diffusion of the primitive heat to which the planets owe their spheroidal form." Acknowledging ignorance of the actual properties of the rocks below the surface, he concluded by analogy that "the inequality of cooling . . . would place the crust under the necessity of continually diminishing its capacity . . . in order that it should not cease to embrace the internal masses exactly" (Elie de Beaumont 1831:264). The suggestion was that the earth underwent periodic readjustments in which the crust diminished its circumference by collapsing (or thrusting up; either would do) along great circles, in order to remain in universal contact with the shrinking interior. These shrinkages resolved long-building stresses in episodes of great violence, causing tidal waves and other disturbances that resulted in widespread extinction of life over the face of the earth.

The theory of long-term cooling and contraction of the terrestrial globe that Elie de Beaumont attributed to Humboldt had been proposed in the seventeenth century by Descartes and Leibniz and in the eighteenth century by Buffon. It had been elaborated in the early nineteenth century as a natural consequence of the Kant-Laplace nebular hypothesis. If the earth and other planets had coalesced from clouds of hot gas rotating around the sun they must, it was reasoned, lacking an independent source of internal heat, be cooling by simple conduction and therefore must be contracting. This contraction could be expressed in some arrangement of fissures, faults, and folds in the outer crust of the earth.

Between 1810 and 1830 a number of attempts were made to establish this theory by observation of the earth's temperature gradient and by model experiments with iron globes. Joseph Fourier (1768–1830) had taken the problem of the earth's heat as one of the principal questions to be solved by his analytical theory of heat (Herivel 1975:197). Fourier developed a theory of secular cooling of the globe, arguing that the figure of the earth and the principles of dynamics could leave no doubt of a high original temperature for the earth, and that the earth's thermal gradient was just what would be expected if the earth was cooling from a hot state (Fourier 1837). Fourier had heated globes of iron and measured the gradient of temperature within, which he found to be equivalent to one degree for thirty meters when the scale was enlarged to the earth's radius. He was aware that it was impossible to compare such work directly with the interior condition of the earth, since the conditions of temperature and pressure together were unknown. However, he felt that iron, with a density close to that of the earth as a whole, gave a correct idea of the gradient, within an order of magnitude (Herivel 1975:202; Grattan-Guinness 1972:ch. 20).[5]

Pierre Louis-Antoine Cordier (1777–1861) also worked on the problem. Cordier, as already noted, had ended the controversy over the origin of basalt with microscopic, chemical, magnetic, and blowpipe analyses which revealed that various basalts and lavas behaved alike and yet quite differently from sedimentary rocks that were very similar in appearance. His work not only helped still the long-standing quarrel between the Huttonians and Wernerians but

5. At the beginning of the nineteenth century the mean density of the earth was known to within 1 percent, the theory of gravitation had been applied to many problems on a rotating earth, there was a close estimate of the earth's ellipticity, and there was some theory on the propagation of disturbances in a deformable medium (Poisson). Bullen gives a concise summary of early measurements of the earth's density (Bullen, 1974:11).

also attacked any interpretation of geologic structures based on external character and macroscopic observation alone (Burke 1971: 411–412). A decade later he collected observations on the temperature gradient in mines and, like Fourier, came to a figure of one degree of temperature for every 30 or 40 meters of depth. He believed that the earth was fluid below 5,000 meters, and that it was a cooled star, consolidating from the outside toward the center (Cordier 1827).

Elie de Beaumont saw his own work in the tradition of Cordier, Poisson, and Fourier—and saw the source and nature of the earth's internal heat as an important geological question. However, he shied away from attempts to give exact estimates of the rate of the earth's cooling or contraction. For this he was often criticized, as were other supporters of the theory of contraction; it was more or less implied that the theory was for that reason wanting in conviction or proof. However, Elie de Beaumont's older contemporary, the Swedish chemist Jöns Jacob Berzelius (1779–1848), supported his empirical caution in refusing to speculate on the rate of cooling or the exact effect of the cooling on the crust of the earth. If anyone in the world had the mathematics for the task, said Berzelius, it was Elie de Beaumont, and it was certainly not ignorance of the way in which such calculations might be performed that stopped him (Frängsmayr 1976:231). It was rather the physical quantities that were missing: the elapsed time since the beginning of cooling, the specific heats of the surface and interior materials, and so on.

While Elie de Beaumont's effort was no *Principles of Geology*, it was an effective resumé of research on the European continent in the tradition of Werner and his followers that could provide viable opposition to the persuasive power of Lyell's uniformitarian scheme of earth history. It rested on a firm foundation in geology of a high caliber and was drawn from the very sciences that Lyell proposed should be applied in the study of the earth—comparative anatomy, botany, mineralogy, chemistry, natural philosophy. Elie de Beaumont accepted erosion and sedimentation, volcanism, plutonic injection, and other aspects of the Lyellian picture of the earth, but he employed them to substantiate a geology that allowed catastrophes and progressive development in the history of the earth. In all, it was precisely the sort of theoretical development that Lyell's progressivist and catastrophist opponents could employ against his implicit claim that the uniformitarian approach and the science of geology should be regarded as synonymous. While it is true that no group of British geologists went over to Elie de Beaumont and began a school based on his work, it is also apparent that recent attention to the theological ramifications of the uniformitarian-catastrophist debate has overwhelmed the strictly geological aspects of the contest between supporters of progressionist and nonprogressionist accounts of earth history, to the latter of which Elie de Beaumont's work was roughly and accidentally parallel. Attention to the second debate will serve to alter vastly the picture we have inherited of the development of geology in the 1830s and 1840s as a rearguard action by uniformitarians against an untenable and theologically motivated catastrophism.

CHAPTER FOUR

The Origin of Mountain Ranges: European Debate, 1830–1874

THE idea of a cooling earth and many other ideas expressed by Elie de Beaumont in 1829 found an influential supporter in England—Henry Thomas De la Beche (1796–1855)—who was soon to become the first director of the Geological Survey of Great Britain. De la Beche was conversant with the geology practiced on the continent: in the 1820s he had read through the back numbers of the *Annales des mines,* and in 1824 had published a series of these memoirs in English to aid in cross-correlation of sedimentary rocks between England and the continent. He also prepared and published synoptic tables of equivalent formations and translated as well Brongniart's classification of rocks. In 1831, as a part of this series of translations, he had arranged publication in the *Philosophical Magazine* of extracts from Elie de Beaumont's lengthy 1829 paper on mountain ranges read before the Académie des Sciences. The publication of this translation was obviously also an element in the controversy then being fought out over the nature and intensity of geological action, which had been spurred by Buckland's *Reliquiae Diluvianae* and Lyell's first volume of *Principles of Geology.* De la Beche was not a diluvialist, nor was he a catastrophist, but he was an opponent of Lyell's severe strictures on the intensity of geological action, especially as it concerned the formation of mountain ranges.

In the first volume of *Principles of Geology* (1830), Lyell had devoted only six pages to the origin of mountains and had discussed only two examples—the Apennines and the Alps. In estab-

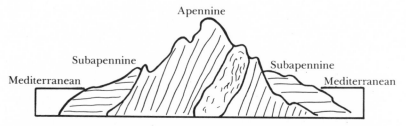

Lyell's cross-section of the Italian peninsula. The extremely primitive and schematic character of the diagram is evidence of Lyell's rudimentary notions of mountain structure and of the purpose of tectonic sections. (After Lyell 1830:I.)

lishing the "principles of geology," Lyell retained Playfair's habit of argument from crucial instances—his theory of climatic change was based on an examination, not of all the known regions of the world, but of the North Atlantic. A similar economy of procedure was used to establish the "principles of geology" in regard to mountain chains. His theory was that earthquakes and volcanism, over long periods, contributed to great inequalities of relief: both high mountains and deep oceans. Notable, however, was the absence of any discussion of the existence of mountain *chains*. His illustration of the Apennines was not a map of their extent but a schematic cross-section of their composition, which relieved Lyell of the responsibility of explaining long stretches of conformable strata uplifted along an axis (Lyell 1830–1833:I, 135).

In discussing Alpine chains, Lyell pictured a form of uplift in which progressive vertical movement, produced by earthquakes, lifted both the core of the chain and the erosion-born strata on its flanks from the surrounding waters. Consistent with his uniformitarian principles, he suggested that the Alps were still rising and would someday be as tall as the Andes (Lyell 1830–1833:I, 137).

Again, he made no attempt to explain the existence of the chain or the deformations of strata. In this sense Lyell's explanation was even less adequate than Hutton's, since Hutton denied any special uplift and accounted for deformation in the period of deposition of the sunken protocontinent. Lyell, once admitting that there were uplifted mountains (apart from continental elevation), was faced with explaining all their characteristics—something he quite clearly was unable to do.

The inadequacy of his account became immediately apparent. In 1831, although "diluvialists" like Adam Sedgwick, convinced by Lyell's criticisms, were abandoning their insistence on the correlation of the Mosaic deluge with the phenomena of geology, they nevertheless remained catastrophists. In their retreat from the deluge, many, Sedgwick included, went no further than the theories of Elie de Beaumont on the catastrophic uplift of mountain chains, which were offered as examples of catastrophic phenomena as a regular part of history of the earth's crust (Sedgwick 1831:311–312).

If they were deprived of their favorite catastrophe, as Charles C. Gillispie put it, they at least retained solid evidence for catastrophes in general (Gillispie 1959:143). The ground of the debates within the Geological Society and the newly founded British Association shifted in 1831, in part, to an examination of the theories of Elie de Beaumont and to whether *they* were conformable with phenomena in Great Britain (Lyell 1830–1833:III, 348). William Conybeare (1787–1857) used his extensive research in stratigraphy to cast doubt on the usefulness of parallelism as a time boundary in England, pointing to deformed series of different age and yet of parallel strike and to others of the same age and different strike. Conybeare's criticism forced Elie de Beaumont into a modification of the details of his plan, and his correlation of the upthrust of the Andes with the biblical deluge was quietly dropped (Boué 1834: 131, 149). When De la Beche's *Manual of Geology* (1831) was translated into French by Brochant de Villiers in 1833, an expanded system of upheavals of mountain chains, now twelve in number, was added in a long appendix prepared by Elie de Beaumont.

Elie de Beaumont's controversial behavior made him as difficult an ally as an opponent. While he read and absorbed criticism of his theory, he never acknowledged its source, nor did he modify his broad theoretical statements, and it soon became evident that retraction was not in his intellectual repertoire. This was particularly galling to his colleagues in the Société Géologique. Ami Boué was a founder of the society and wished to work toward a reconciliation of the positions of Lyell and Elie de Beaumont, which, he thought, were not as far apart as others claimed. He agreed with Elie de Beaumont's theory that mountain upheaval was a distinct and rapid geological phenomenon, he disagreed, as did most of the geological community, with Lyell's idea that it was accomplished by "unlimited repetition of local and slow movements" (Boué 1834:123). He also accepted with Cordier and Elie de Beaumont the cosmologically deduced idea of *refroidissement séculaire* ("long-term cooling") against the doctrine of plutonic uniformity—Lyell's steady-state earth. Yet he was attracted to the general tenor of Lyell's work and thought it highly valuable for geology; he is generally considered to have been, with Constant Prévost, one of Lyell's strongest supporters in France.

Boué's attempt to influence Elie de Beaumont into a modification of the grand scheme of mountain upheavals was singularly unsuccessful. Reviewing the revised system of twelve upheavals published with the translation of De la Beche into French, he lamented that Elie de Beaumont was sacrificing the cogency of his system to his pride of originality, and in response to Boué's and Conybeare's demonstration of asynchronous parallelism, had merely multiplied the number of upheavals to account for it (Boué 1834: 134). Moreover, in the details of the system, Elie de Beaumont had acknowledged that the number of revolutions was perhaps not fixed, that parallelism was in fact not absolute, and that sometimes orogeny recurred in the same direction—all without any modification of the original (and now misleading) generalizations that prefaced the 1831 essay (Boué 1834:130, 132). All this put Boué in a difficult position. He supported the substance of Elie de Beaumont's account of mountain genesis, particularly the idea of the Alps as a complex and multiple uplift and not a single upthrust, and also the general idea that periods of relative tranquillity and activity alternated in geological history. But he was at the same time repelled and irritated by the posturing of the grand theorist.

Lyell, however, was also unmoved and unmovable in the matter of general theory. The third volume of *Principles of Geology* was written in the course of the debate, and a full chapter was devoted to an examination and criticism of Elie de Beaumont's general theory. Lyell marshaled all the evidence he could find against the proposition that mountains were catastrophically upheaved in the manner suggested by Elie de Beaumont, and he disputed every part of Elie de Beaumont's assertion that parallel chains had been uplifted simultaneously and intercontinentally. The critical opposition to uniformitarian theory was the postulation of cataclysmic uplifts that punctuated the secular processes with episodes of great violence. Lyell admitted that Elie de Beaumont was "eminently qualified" to undertake an attempt to reduce observations on the elevation of parts of the continents to a "systematic whole" and in so doing had demonstrated a "considerable knowledge of the facts." However, this factual mastery was combined with an "ardent love of generalization," which had led him into error (Lyell 1830–1833: III, 337). Lyell professed no objection to the dating of uplift from unconformities on the flanks of mountain chains, as long as individual strata were not confounded with the periods to which they belonged, and as long as "due latitude" was given to the term *contemporaneous*. But this, he insisted, was exactly the confusion into which Elie de Beaumont had fallen. Failing to distinguish between formations and the periods in which they had been laid down, he had developed the idea of convulsive uplift. However, Lyell objected, knowledge of the period in which the horizontal strata at the base of the chain were laid down was no index of the time in which the tilted strata were upraised. A long period of erosion might well have intervened between the two depositions, and the possibility of such an interval destroyed, in Lyell's mind, the claim for convulsive contemporaneity (Lyell 1830–1833:III, 341). Even if it were admitted that two ranges, roughly parallel, exhibited the same sedimentary sequence at their bases, this was no guarantee of contemporaneity unless the word were very loosely construed; *any* two ranges might have originated between the deposition of two formations.

A second problem with Elie de Beaumont's theory, from Lyell's uniformitarian standpoint, was the assertion that mountains were uplifted by a process other than volcanism and earthquakes, which in Lyell's theory provided the small increments of uplift that eventually resulted in great inequalities of relief. Elie de Beaumont had argued that the conformable disruption of strata for long distances was evidence of a concentrated episode of mountain building that had split the strata apart. This assertion was compatible both with Buch's theory of catastrophic plutonic uplift and with the theory of catastrophic adjustment of the crust to a shrinking and cooling core, toward which Elie de Beaumont had leaned in his memoir of 1829. Lyell objected that in extrapolating his theory from the mountains of Europe to distant chains "never touched, as M. Boué remarks, by the hammer of a geologist" Elie de Beaumont had made the unwarranted assumption that "in these distant chains, the geological and geographical axes always coincide," whereas it was well known that in Europe the strike of the beds is not always parallel to the direction of the chain (Lyell 1830–1833:III, 346–347).

These were cogent and well-received criticisms and were supported by many geologists not at all identified with Lyell's uniformitarianism: Henry De la Beche, William Conybeare, L. A. Necker de Saussure, and Bernhard Studer had all objected either to one or another part of Elie de Beaumont's theory (Lyell 1830–1833:III, 347).

What is most striking about these arguments, however, is not what they were able to achieve against the notion of catastrophic uplift, which was substantial, but what they forced Lyell to concede against his own principle of uniformitarianism. Lyell did not propose any other theory of mountains in later volumes of *Principles of Geology* than that proposed in the first volume. Though he showed that not every chain exhibited a continuity of strike of the strata with the strike of the chain, betokening some homogeneous uplift, he tacitly admitted that many ranges did show such a parallelism. Further, even if absolute contemporaneity of uplifts could not be demonstrably shown, the consistency of certain unconformities provided ample room for argumentation that some periods in the history of the earth showed a greater tendency toward the creation of mountain chains than others, and thus his strict interpretation of uniformitarianism became untenable. If the regularity of mountain formation was less than Elie de Beaumont had desired to prove, it was measurably greater than Lyell's theory could allow. Finally, Lyell avoided discussing the association of mountain-building episodes with the theory of the cooling of the earth, while noting that Elie de Beaumont had asserted that such an association was possible (Lyell 1830–1833:III, 339). In the first volume of *Principles of Geology*, Lyell, with typical reserve, had not rejected the idea of the cooling of the globe but had asked that any theorizing based on its assumption wait for a study that would establish the existence of the earth's central heat, its distribution, and its secular variation (Lyell 1830–1833:I, 143). In the meantime, he thought it shared with other cosmogonical notions the failing that "by referring the mind directly to the beginning of things it requires no support from observations, nor from any ulterior hypothesis" (Lyell 1830–1833:I, 104–105). This was the same argument that he had offered against the invocation of "ancient causes" in shaping the earth, causes which could not be determined by a study of the earth's present condition: "Never," he wrote, "was there a dogma more calculated to foster indolence and blunt the keen edge of curiosity" (Lyell 1830–1833:I, 2).

De la Beche responded for the opposition in 1834. Like Boué, he found Elie de Beaumont's grand correlations extreme, but his *Researches in Theoretical Geology* (1834) reveals his general attraction to Elie de Beaumont's approach. Earlier in the year he had heard that Elie de Beaumont had used Brochant de Villiers's translation of his *Manual of Geology* to offer a revised version of his theory of mountain upheavals. Not having seen the translation he could not endorse the doctrine but argued that

What lines [of mountain uplift] may eventually be found to prevail will, as previously noted, require much time and great pa-

tience to discover; but . . . geologists will not the less have reason
to feel thankful to M. Elie de Beaumont for having rescued the
subject from the state in which he found it; it being impossible but
that the investigations which this theory will necessarily give rise,
must end in the most important additions to geological knowledge. [De la Beche 1835:490]

In *Researches in Theoretical Geology*, a work overshadowed historically
by Lyell's *Principles of Geology*, De la Beche cautiously pursued a
middle ground between uniformitarianism and the theory of paroxysmal upheavals, and in so doing advanced many of Elie de
Beaumont's proposals in a form palatable to geologists who found
Lyell's earth too quiescent and Elie de Beaumont's too violent.

De la Beche was sympathetic to the theory of a hot interior for
the earth and its "former igneous fluidity" because it provided the
simplest explanation of a host of observable facts. But he held it
only as a hypothesis and would adhere to it, he said, only as long as
it remained the simplest explanation of those facts (De la Beche
1834:v–vi). The first chapters of *Researches* dealt with cosmogonic
questions, the densities of the planets, the figure of the earth, the
elemental contents of the earth's crust, geochemical phenomena,
and other physical topics. The next chapters dealt sympathetically
with geological agencies important in the Lyellian scheme and accepted completely the idea of geological time as very extensive
indeed.

When he turned to an examination of mountain ranges he
parted company with Lyell—the fragmenting of the crust in mountainous areas was a most "unusual phenomenon" and required some
special explanation. Yet he also explicitly disagreed with Elie de
Beaumont on the geographical extent of chain systems along great
circles, and on the strength of the forces of uplift. He favored
more local uplift, along distances much less than those proposed by
the French geologist (De la Beche 1834:119–120).

Nevertheless, he was strongly attracted to Elie de Beaumont's
application of the theory of the cooling of the earth as an explanation of the genesis of mountains in clearly defined linear zones. "If
we suppose," he wrote, "with M. Elie de Beaumont that the state of
our globe is such that, in a given time, the temperature of the
interior is lowered by a much greater quantity than on its surface,
the solid crust would break up to accommodate itself to the internal
mass; almost imperceptibly when time and the mass of the earth
are taken into account, but by considerable dislocations according
to our general ideas on such subjects." Since it was likely that
new dislocations would most easily be effected along old fractures,

"under favorable circumstances, broken and tilted masses would be thrust up into ridges or mountain chains" (De la Beche 1834:121). Lyell was not uncomfortable with a theory of mountain building based on successive dislocations by earthquakes, but De la Beche was extremely skeptical about an accumulation of small forces equaling the effect of sudden exertions of greater power. Modern earthquakes, on which Lyell based his dynamical hopes, showed nothing comparable with the great dislocations observable in mountains. While movement along faults could be caused, of course, by continual small forces, the contortions of the strata required some greater explanation. Here De la Beche invoked James Hall's experimental simulation of contortion by lateral pressure. Hall had constructed a "squeeze-box," a rectangular trough filled with layered sand and mud, and had screwed in the ends by means of a vise, producing contortions in the layered material similar in every way to the contorted strata observed in mountains (Hall 1815:79). De la Beche accepted this as suggestive evidence that contorted rocks had once been in a yielding state and subject to forces of considerable intensity. Since mountain chains like the Alps and others that had figured in Elie de Beaumont's theory showed every variety of folding, faulting, and dislocation, mere uplift by earthquakes of present intensity was inadequate to explain their genesis.

Those who derided mountain elevation by forces greater than earthquakes were, in his mind, to be compared with those who would scoff at the expense of effort of using a pile driver, "assuming a man atop the pile with a hammer could drive it as well" (De la Beche 1834:126). Lyell's account, he said, was insufficient to explain the magnitude of dislocation in mountains, and it also failed to give a reasonable account of their structure. The symmetric contortions of strata, in the Central Alps in particular, seemed to indicate a very concentrated force from the center up and outward. De la Beche concurred with Elie de Beaumont that the structure of mountain ranges suggested a directed thrust of some sort and not accumulated episodes of generalized "uplift." Such directed thrust could be explained alike by central elevation of the chain or by a peripheral subsidence, which would provide a lateral pressure on strata heated and saturated by water. The strata, weighted from above, would not be able to escape contortion under such pressure and would be simultaneously compressed and elevated (De la Beche 1834:138).

In either case, however, the dislocations were of a magnitude and expressed tendency at once inadmissible and unexplained within the confines of the system of geology proposed by Lyell. De la Beche's iteration of favored themes among European geologists—concentrated uplift, extensive contortion, inclusion of the total history of the earth as a planetary body in explanation of regularly occurring dynamical phenomena—overrode Lyell's specific and revisionary criticisms of Elie de Beaumont's overall scheme. However, De la Beche, Elie de Beaumont, and the British progressionists were unable to make common cause against Lyell and his supporters at this time. The very heterogeneity of the opposition against Lyell's version of earth history served to advance it; agreed in criticism of his approach, the opponents of secular change by present agencies could not agree on a response. From the complexity of motives that led them to oppose Lyell, no single alternative version of the history of the earth arose in England to rival the consistency that was the strength of Lyell's *Principles of Geology*.

This is the more evident because at the moment when De la Beche was defending his theory in *Researches in Theoretical Geology*, Elie de Beaumont was turning away from the very aspect of it that had recommended him to the British geologist—the theory of secular contraction as an explanation for the upthrust of mountain masses. In 1833, when Lyell had met Elie de Beaumont in Paris, he noted that the French geologist was already trying to unite Buch's work on "elevation craters" with his own more general theory of uplift, and Lyell interpreted this move as an attempt to strengthen the theory of mountain uplift in the face of his own telling criticisms laid out in the third volume of *Principles of Geology* (Wilson 1972:383). With both sides in the dispute now addressing the character of plutonism—the uplifting force of molten matter pressing from below—the debate shifted to the manner in which this uplifting force was expressed; generally, in continental uplift, or specifically, in isolated and axial outbursts at particular points in the crust.

In 1829, Elie de Beaumont had wavered toward the theory of Buch and then rejected it. By 1834, however, the idea was once again attractive to him. He traveled to examine the volcanic chimneys (*puys*) of central France and reported that they seemed to support Buch's idea of elevation craters (*Erhebungscentra*), which had recently declined in popularity, no doubt partly as a result of Elie de Beaumont's own writings (Elie de Beaumont 1833). In 1835 he accompanied Buch on a geological excursion to Mount Etna and was convinced that his theory served well to explain the phenomena surrounding volcanic cones (Sainte-Claire Deville 1878:351). In this theory, which has still to find a sympathetic account in the history of geology, it was asserted that the cones of volcanoes existed because the pressure of plutonic rock from below first forced up the strata radially around a center of elevation and then broke through at the apex, covering the uplifted strata with lava. At first reserved about the application of the theory to mountain ranges Elie de Beaumont became more and more enamored of this interpretation of plutonic forces, and this carried him further and further from the idea of subsidence and compression as a cause of elevation, which De la Beche had found so felicitous in the context of the theory of secular contraction.

On the way home from a meeting of the Institut de France in Paris in 1835, Lyell talked with Buch and Elie de Beaumont, who told him that they vociferously opposed his attempt to explain uplift in terms of regional elevation and that they were planning to publish to that effect. Also at this time Lyell was developing the "drift theory" of the transport of erratic boulders by ice, and he was employing the apparently sound evidence of a gradual uplift of Scandinavia, measured by the tide markers along the Baltic coast, in order to oppose notions of a catastrophic elevation of any kind. The cooling of relations between Lyell and Elie de Beaumont is evident in his portrayal of the conversation. Buch grew passionate and argued that he would fight Lyell's ideas "to the death." "De Beaumont," wrote Lyell, "voiced an obsequious assent" (Wilson 1972:414).

The theoretical point at issue in the debate over "craters of elevation," which lasted from 1834 to the middle 1850s and pitted Buch, Humboldt, Elie de Beaumont, and Charles Sainte-Claire Deville against Lyell, George Poulett Scrope, and Constant Prévost, was the meaning of *volcanism*. The question was posed thus by Zittel: "Is Vulcanism an active agent, or a passive accompaniment of regional elevations?" (Zittel 1901:265). There was no avoiding the controversy, which turned on a number of disputed points, including the petrology of igneous and plutonic rocks. From the standpoint of the late nineteenth century, the cogency of Buch's arguments is revealed in Zittel's opinion that Gilbert's work on the "laccolites" of the Henry Mountains of Utah in the 1870s was "reminiscent of von Buch's Elevation-Craters" (Zittel 1901:274). In the contemporary context, Elie de Beaumont's employment of the theory was, as Lyell suspected, a way to shore up his position against Lyell, with the aid of Buch's reputation and massive learning.

The irruption of the debate called forth an anxious response from the French volcanologist Constant Prévost, another founder of the Société Géologique. Prévost had made geological excursions with Lyell and, like Boué, admired his work, but also like Boué he was a prominent supporter of Elie de Beaumont's geometrical analysis of the intersection of mountain chains. He was therefore appalled to see the widening rift in geological theory signaled by the increasing attachment of Elie de Beaumont to a theory of absolute elevation of mountains allied to the elevation-crater hypothesis. He argued, as had De la Beche, that the attribution of surface relief to one or another version of the theory of subterranean power was a matter of simple preference. Given a point A, higher than B and C on either side of it, how could one definitely assert that A had risen when it was equally cogent to argue that B and C had subsided (Prévost 1840:186)? Prévost took special pains to deter Elie de Beaumont from Buch's analyses of volcanoes. One of the arguments that had convinced Elie de Beaumont of the soundness of the theory of elevation craters was the presence of marine shells on the flanks of the cone of Etna, which seemed to show that the cone was a radial uplift of strata of marine origin. Prévost, who had made careful studies of volcanoes himself, responded that it was not uncommon for volcanic ejecta to contain portions of the "country rock" through which the volcanic chimney passed, and which had been expelled during eruptions. Prévost had long been attached to the idea of subsidence following secular cooling as a cause of surface relief, and it is perhaps to him that we owe the earliest use of the "drying apple" analogy. The earth, constantly cooling, would in this analogy be like an apple taken from the oven—the surface would at first be smooth but would soon become corrugated as it cooled.

In 1840, writing against the hypothesis of absolute uplift, he criticized the misunderstanding of the earth's mechanism implied in the use of the word *uplift*. If we are to call the wrinkles on a drying apple "uplifted," he wrote, or say that the sap oozing from a tree is "uplifted," or that the head on a glass of beer is "uplifted" from the bottom of the glass, then further struggle against the term is useless. However, he argued, there was no need to speak of "uplift" in the ordinary sense at all when discussing the creation of structure or relief (Prévost 1840:186). On the other hand, Prévost was unable to marshal arguments that could establish the reality of merely apparent uplift, actually due to general peripheral subsidences in a sinking crust, and thus could not moderate the dispute; in any conflict, the voice of moderation is the first to be stilled. The stridency of competing factions fills the air (and the journals), and the moderates nervously await the exhaustion of extreme positions. Yet in Prévost's plaintive cry for toleration and agnosticism, there is a failure to acknowledge the driving power of a systematic overview, a general theory, that motivated both Lyell and Elie de Beaumont to continue the struggle and to gather support where they could find it to advance their central conceptions. Somehow Prévost's stance rings false, as if theory could develop without advocacy in the patient accumulation of observations.

Lyell, like Elie de Beaumont, searched for advocates who might advance the cause of his theory. Although outnumbered by his opponents and unable to convince the moderates like De la Beche, Boué, and Prévost that his theory of regional uplift was adequate to explain mountains systems, he did not stand alone in the debate over the rate and nature of geological activity. He was soon joined by John Herschel (1792–1871) and by Charles Babbage (1792–1871), Lucasian Professor of Mathematics in Cambridge University. Both men accepted the Huttonian time scale, the primacy of the erosion cycle, and the slow and regular course of geologic change. They both subscribed to the notion that deformation of strata was not the result of any catastrophe of uplift but something that happened to the strata while deeply buried and exposed to the joint action of great temperature and pressure. They accepted the idea of Hutton and Lyell that volcanoes were a kind of safety valve for the earth that periodically allowed the escape of subterranean pressure and heat, thus preventing greater and more general disruptive cataclysms, and therefore were not agents of uplift.

Of the two, Babbage was the more committed to the establishment of Lyell's uniformitarian system and the more active participant in the debates. In 1837, Babbage published *The Ninth Bridgewater Treatise: A Fragment* (Babbage 1837). The work was an independent and uninvited addition to a series of eight treatises published between 1833 and 1836 that had been composed to show the beneficent activity of the Creator in the ordering of various parts of the natural world. These "Bridgewater treatises," established by a bequest of the late Earl of Bridgewater in 1829, were important texts for the physicotheologians who sought to demonstrate the participation of the Deity in the ongoing course of nature. Written by scientists in many fields, they were widely and enthusiastically read and passed through numerous editions, some published as late as the 1880s (Gillispie 1959:209ff). The eighth, and last to appear, was William Buckland's *Geology and Mineralogy, Considered with Ref-*

erence to Natural Theology (1836). Buckland, it will be remembered, was one of the most prominent geologists in Great Britain and the author of *Reliquae Diluvianae* (1824), in which he had argued that many of the phenomena of geology testified to the existence of a former universal deluge. He was one of the foremost of the diluvial geologists and catastrophists and an opponent of Lyell and the uniformitarian system.

Babbage's *Ninth Bridgewater Treatise* argued for a uniformity of nature not on such a foundation of divine aid and intervention but according to simple and regular natural laws that determined the phenomena without the constant attendance of the Deity. Babbage's principal target was the philosopher William Whewell, author of another of the Bridgewater treatises, *Astronomy and General Physics, Considered with Reference to Natural Theology* (1836). Although Whewell had been more moderate in his claims about the intensity of the Lord's ongoing activities than some of the other authors, the thrust of the work was to demonstrate that the regularity of celestial and physical phenomena testified to a governing intelligence and to an order in the universe not discernible by physical investigation or mathematical proof. The major part of Babbage's treatise was a response to this claim on the grounds already noted, but he appended to it some remarks that bore directly on geology and on the adequacy of the Lyellian system. Lyell was opposed to the invocation of earth contraction and secular cooling to explain the phenomena of geology and preferred to assume that the earth had an inexhaustible supply of internal heat. However, his own treatment of the causes of uplift and subsidence and the relation of the crust to the interior of the earth was vague and unsatisfactory by the standards of the time and had left him without an adequate explanation of the origin of mountains. Babbage sought to improve and expand upon the theory of uplift and subsidence presented by Lyell in *Principles of Geology* and in so doing to remove this major source of criticism of the uniformitarian theory.

He was already at work on his version of the system of general uplifts and subsidences when he received a copy of a letter from Herschel to Lyell that considered the same problem. In 1834, Lyell had written to Herschel, then in South Africa, with a number of questions about aspects of the system of the world, and Herschel's response contained some extremely ingenious suggestions about the way in which erosion and deposition, uplift and subsidence, earthquakes and volcanoes, the deformation of strata, and even the regularity of mountain ranges and their locations could all be seen as complementary aspects of a single process. Babbage and Lyell were anxious above all to establish the phenomena of geology with reference to a few simple processes, and Herschel's ideas seemed a strong argument for the reasonableness of their position. Babbage therefore included Herschel's letter as an additional appendix to *The Ninth Bridgewater Treatise,* with a glowing commendation (Herschel 1837).

Herschel was concerned above all to provide that "great desideratum of the Huttonian theory," a source of internal heat to occasion the melting and deformation of strata, and an expansive force which might give a reasonable explanation of the uplift of the continents. Herschel, even more than Babbage, was dissatisfied with Lyell's vague notion of circulating currents of hot matter in the substratum. Like many of his contemporaries he was interested in possible connections between the earth's magnetic field, electricity, and the earth's internal heat, as they might conspire to create geologic structure. Necker de Saussure, De la Beche, James Dana, and others had toyed with what little they knew of these matters to add to their meager fund of information about the interior of the earth (Dana 1847b:393).[1] The most promising line of contemporary research was the attempt by numerous observers to establish the earth's temperature gradient. It had been shown that temperature increased with depth, but estimates of the rate of increase and the depth at which melting would consequently occur varied widely; calculations of the latter were further complicated by ignorance of the effects of pressure and temperature together on matter at various depths.

Herschel was among those convinced that whatever the actual gradient, the earth achieved a temperature within a few kilometers of the surface high enough to rob the rocks of their strength to resist deformation. Upon this postulation he built what he thought was a plausible mechanism for explaining the entire range of geological activity *within the sedimentary cycle alone.* Without any further stipulation about the earth's interior than that it was hot, without reference to its thermal history or to any gratuitous circulations of

1. L. A. Necker had noticed that the lines of equal intensity on Sabine's geomagnetic map seemed to follow continental coasts and the strike of major mountain chains. Dana cites Necker's speculations in the *Bibliothèque universelle des sciences et des arts de Genève* for 1833, p. 180, but Necker's original article, "Sur quelques rapports entre la direction générale de la stratification et celles des lignes d'égale intensité Magnetique dans l'hemisphere boréal," appeared in the *Bibliothèque universelle* in 1830.

hot matter or magnetic or electric forces at work beneath the crust, he devised an explanation that reduced all geological activity to erosion, deposition, and heat (Herschel 1837:213).

Herschel chose as his example of this pleasing natural economy the uplift of Scandinavia. Tidemarks along the shores of the Gulf of Bothnia (between Sweden and Finland) had shown a systematic drop in the level of the sea, verified in a series of measurements over more than a century. A debate was already raging over the meaning of this change in level, which had amounted to almost a meter in a hundred years. If the ocean level were dropping, this would be evidence for the theory of secular contraction of the earth, which postulated deepening ocean basins. If the bordering lands were rising slowly and regularly, this would serve to confirm the idea of continental uplift and Lyell's contention that the repair of the land proceeded even as it eroded away. Today the elevation of the land—the "fennoscandian uplift"—is explained as the elastic rebound of the crust of the earth following removal of the great continental icesheets of the last glaciation, an idea first put forth in 1862 (Bailey 1935:19).

Proponents of the theory that the land was rising pointed to the existence of ancient strandlines (former beaches), above the present shoreline, sometimes hundreds of meters above it. Darwin had argued for the gradual uplift of the Andes after examining successions of such displacements of the strand along the coast of South America. Of course, while a number of successive raised beaches are a fair argument against catastrophic and instantaneous elevation, they are no argument against the theory that the sea level is dropping—the former beach will be displaced upward from sea level whether the water falls or the land rises.

The question was resolved in Herschel's approach to the matter—in his hypothesis the elevation of the land and the sinking of the floor of the ocean were simultaneous aspects of a single process, though in neither case was the direction of movement permanent. Herschel reasoned that as the continental platform was eroded by rain, frost, and running water, the products of erosion were carried to the sea where they were deposited as new sedimentary beds. Because these water-filled sediments had a density greater than water alone, they constantly increased the load on the seabed below. Since, in his view, the seabed rested on some deeper semifluid or mixed mass whose composition was unknown but whose temperature could be inferred to be sufficiently high to render it mobile, the constantly increasing weight of the sediments had to

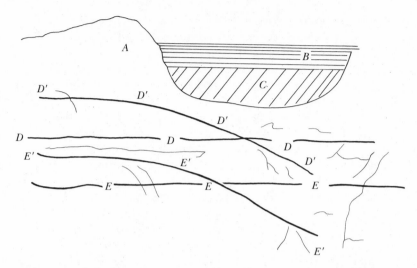

John Herschel's picture of the uplift of Scandinavia, 1834. *A* is Scandinavia, *B* the Gulf of Finland, *C* recently deposited sediments, *D* is the seabed, *E* the semifluid matter beneath. The weight of the recent sediments (*C*) depresses the seabed (*D*) to *D'* and displaces the semifluid matter (*E*) to *E'*, which elevates the shoreline (*A*) and renews the erosion gradient. (After a drawing in Babbage 1837: app. G.)

depress the ocean floor, which forced out laterally the mobile matter beneath (Herschel 1837:213). This lateral motion produced a steady elevation of the land as the mobile matter passed underneath the continental crust of Scandinavia; the rate of this change in elevation was measured by the tidemarks.

Herschel's work says nothing about glaciation, of course, nor is there consideration of the relative densities of the seabed and the sediments and the way this might affect the amount of load without plastic deformation. Similarly untreated are the probable distribution of sediments in an ocean basin and other considerations pertinent from the modern point of view. But it is probably the very simplicity of the plan that made Herschel's idea so long-lived and influential. Its basic outline was evident sixty or seventy years later in the *Unterströmung* ("underflow") theory of Otto Ampferer and the theory of isostatic compensation of Clarence Dutton, who acknowledged the priority of Herschel (and Babbage) in presenting the theory of dynamic readjustment at continental margins through loading (Dutton 1889/1925:364).

One of the outstanding advantages of the hypothesis was that it linked the erosion cycle causally to the uplift and subsidence of different portions of the crust, as a kind of motor, perpetually preventing the exhaustion of relief and the planation of the earth to sea level. For Herschel, "every continent deposited has a propensity to rise again, and the destructive principle is continually counterbalanced by a reorganizing principle beneath." With the erosion cycle above and the "inexhaustible supply of heat from below," the circuit could be indefinitely maintained (Herschel 1837: 213). Also, the same process explained the occurrence of volcanoes at the margins of continents (e.g., the Pacific coast of South America, the Mediterranean). Accepting the notion of volcanic activity as the release of gas or fluid pressure, he thought that the continual elevation and loading of the continental edge would lead eventually to a break of the "lateral supports at or near the coastlines," opening a chain of volcanoes as pressure escaped through the faults and fissures created by the ruptures (Herschel 1837:213).

Thus far the hypothesis was a contribution to the theory of the mobility of the crust, of continental elevation and oceanic depression. However, in the Lyellian scheme, as in the Huttonian, an answer had to be found as to how the sediments from the ocean floor were raised to become a new continent, and their new appearance—contorted, deformed, often recrystallized, interbedded with and invaded by lavas and plutonic rocks—had to be explained. The catastrophists and the exponents of the theory of secular cooling and contraction explained these phenomena as resulting from the upthrust of molten matter, simultaneous deformation and invasion in cataclysmic episodes, transverse compression and rupture with attendant volcanism, and various other schemes opposed to the Huttonian idea that deformation occurred at depth and that whole continents were raised and then eroded to expose preexisting deformation and alteration. Herschel had gone some distance toward explaining the creation of mountains and deformation at the margins of continents, but the problem still remained of explaining how sediments on the floor of the ocean could get close enough to the earth's internal heat to be deformed by it, and in general, with the mobile matter constantly flowing away beneath, how the cycle might ever be reversed and the floor of the sea raised into a continent.

These were the problems that Babbage most eagerly sought to overcome for the Lyellian theory; in particular, the problem of getting the sediments on the ocean floor close to the heat of the earth's interior. Babbage had assumed that the isogeotherms (lines of equal temperature under the earth) would necessarily be depressed under the ocean, for otherwise the bottom of the ocean must, against common sense, be close to a source of great heat; thus the strata would never get close enough to the internal heat to be fused by it (Babbage 1837:191). Here again Herschel had an idea. He theorized that the isogeotherms were spherical from the center of the earth outward, the temperature decreasing uniformly toward the surface. This assumed that the earth was homogeneous and isotropic at each level within, and given that assumption the postulation was logical no matter what the actual gradient might be. At the upper boundary (the bottom of the continental crust and the seabed) the isogeotherms would cease to be spherical and would follow the irregularities—the swells and depressions—of the bottom of the crust. However, whereas Babbage had thought that the internal heat of the earth would be further and further removed from the upper layers as sediments were deposited, Herschel argued the contrary. As a depression in the crust became filled with sediment, as on the bottom of the ocean, the surface of the crust would be smoothed out and the isotherms would rise into the sediments. With this rise, the most deeply buried sediments would begin to melt, and under the increasing pressure caused by new depositions, the lower layers would give way, allowing steam, hot vapors, and lavas to escape upward with the desired deforming and metamorphic effects on the superincumbent layers (Herschel 1837:209).

For Herschel, two complementary processes were at work. The removal of matter from the continents to the floor of the oceans destroyed an equilibrium of *pressure* and led to continental uplift, oceanic depression, and faulting and volcanism at the margins of the continents. The filling in of the oceanic depressions in the crust destroyed an equilibrium of *temperature,* as the isotherms rose into the thickly deposited and deeply buried strata, and caused melting, deformation, metamorphism, submarine volcanism, and the creation of dikes and veins.

Following Herschel, Babbage schematized this process in a way that made it point for point more conformable with the Lyellian theory. But on a matter equally pressing, Herschel had provided him no aid, and it was left to Babbage to devise an answer to the all-important question of how the buried strata were to be reelevated and become continents. Herschel had stated that the movement was not continuous—the oceans could not sink indefinitely toward the center of the earth, nor could the continents rise forever. But while he had shown that the process might stop, he had not suggested a way in which it might be reversed.

Babbage's solution was not a very great improvement over Lyell's and actually ran counter to certain elements in Herschel's pleasing synthesis. Babbage had made room for new strata on the floor of the ocean, not with Herschel's dynamic mechanism, but by supposing that the strata at the deepest level would contract as they became heated, much in the way that mud and clay contract in the bed of a dry lake. This contraction of the lowest layers would allow more room for strata above as the volume decreased at the bottom of the pile. Babbage seems to have believed that rather than destroying the lowest strata, the heat would harden them, making a barrier behind which the heat of the substratum might collect and eventually raise the entire mass far enough that evaporation and the drying of the ocean could be the first step toward the creation of the new continent in place of the old ocean. However, the expense of suggesting this mode of uplift for the whole continent was a contradiction of the process for metamorphism and deformation suggested by Herschel—the heat from the oceanic floor could not both be collected and stored and be simultaneously released to do the work of altering sediment. At any rate, Babbage's solution was little more than a restatement of the original postulation—somehow the heat of the earth raised continents from the ocean floor.

These arguments, particularly Herschel's, did have one great point in their favor: they kept the earth constantly swelling at the surface (as Lyell required) "in spite of the doctrines of refrigeration and contraction" (Herschel 1837:213). Indeed, there is a natural antipathy between Herschel's idea and the idea of contraction when they are applied to the globe as a whole, and the late-nineteenth-century debate between the proponents of the contraction theory in its "classical" form, and the advocates of the theory of isostasy (developed on a foundation of Herschel's idea) played out the antagonistic consequences of the two postulations. Contraction theory depended on the foundering of continental fragments into the interior of the earth—it was a theory of progressive subsidence of the crust. Any doctrine that tended to preserve topographic profiles with dynamic uplift as a *consequence* of such subsidences was necessarily radically opposed to it. Immediately, however, the thrust of Herschel's work was to moot the question of secular contraction: even if it were true, as many of his contemporaries urged, it in no way detracted from the dynamic interplay he had suggested.

Whereas the contractionists urged that a strong crust must give way in a catastrophic adjustment to a shrinking interior, he suggested a more gentle accommodation based on the idea that the crust, on its lower side, was flexible and was constantly and imperceptibly adjusting itself to the behavior of the molten matter beneath, here rising and there sinking in a dynamic equilibration.

This aid from Babbage and Herschel strengthened Lyell's position, since it united opposition to the hypothesis of elevation craters with a quite neat and sophisticated explanation of how the earth's internal heat might produce volcanism as a consequence of regional elevation, well within the confines of his own uniformitarian theory, and even as a result of it. Even so, Lyell lagged far behind his European opponents in understanding the complexities of particular mountain ranges, a failing he wished to correct. However, his first approach to the mysteries of Alpine structure left him baffled. In 1835 he made an assault on the structure of the region explored and mapped by the Swiss geologist Bernhard Studer, and wrote to a friend that

> I found when I attempted to understand the geology of the neighborhood of the lake of Thun, even with Studer's published book of maps and sections, as he calls them, in my hand, that I could not at all comprehend it, nor make out what he meant by his numerous formations. I therefore determined to make myself master if possible of the geology of this part of Switzerland, on which more has now been written than any other part of the Alps, before I made an attack on the less known districts. This I have in some measure accomplished. [Wilson 1972:410]

Lyell, in short, was unprepared for the tectonic complexities that were the substance of the theory put forward against him by Elie de Beaumont, and like many a geologist before or since, he discovered that what looks neat and resolved in the written report can be most puzzling in the field, especially in such highly deformed districts. Moreover, and quite tellingly, his bafflement implies that Lyell was unfamiliar with the concept of the tectonic section, a standard interpretive technique employed by continental geologists since before he was born. It is no wonder that his mountain theory was suspect and that even geologists who supported him elsewhere were disinclined to join him in attributing the genesis of great fold mountains to regional uplift.

In France, the Lyell-Herschel system was never seriously entertained as an explanation for mountain genesis. Later, in 1850,

Constant Prévost undertook a historical review of the question that had "only too frequently" occupied the Académie des Sciences since 1835; the question was whether mountains came about by uplift or by subsidence (Prévost 1850:440). Technically Lyell's theory might have found a place with the proponents of uplift, but it did not. One suspects that the main reason for this is that regional uplift explained the relief of mountains without giving a key to their structure, while the craters-of-elevation hypothesis did: Herschel's model of coastal uplift and simultaneous, parallel subsidence with volcanism and deformation seemed to apply to the Andes, which were far away, and not to the Alps, which were present, and in general could not explain the variety of mountain types nor their orientation to the sea, which in Europe was as often perpendicular as parallel to the coast. For Prévost, the contest concerned on the one hand those geologists who invoked "extraordinary causes" and believed that the rupture, convolution, and realignment of the strata were the result of the *periodically* active expansion of subterranean forces, which, in their efforts to escape, ruptured and heaved up the superincumbent mass, and on the other hand those who referred surface relief to the laws of cooling (*lois de refroidissement*), "a natural consequence of the shrinking and folding of the consolidated envelope of the terrestrial spheroid." That there was still a debate at all was largely Prévost's doing; his opposition to uplift theory had been tireless for thirty years, ever since he had read Jean André Deluc's writing on contraction and had made his own studies of volcanic islands. He was convinced that the total amount of subsidence was greater than the amount of uplift, but he also admitted that he still had only doubts about and no sound arguments against the "seductive hypothesis of the celebrated Prussian geologist" (that is, Buch's theory of elevations). Perhaps, Prévost urged, the time had come to drop the question of uplift and subsidence entirely and to speak only of dislocation without prejudgment of the direction of vertical motion (Prévost 1850:441–443).

One of the reasons for the timing of this appeal was the recent appearance (that same year, 1850) of a paper by Elie de Beaumont in which he abandoned his advocacy of the theory of uplift, held since his visits to Italy with Buch in 1834. The paper included a much revised and extended version of the geometrical theory of mountains in which he was no longer concerned to establish the *exact* nature of the force that provided the movement, but in which he once again emphasized the transverse (lateral) crushing of mountain strata attendant on the secular contraction of the globe (Elie de

Beaumont 1850). This was the very direction in which De la Beche had extended the theory some years before, and he had continued to develop the theory with new observations. In 1846, in the first volume of *Memoirs of the Geological Survey of Great Britain*, De la Beche had described the folding of the mountains of South Wales as "adaptation to a complicated lateral pressure" (De la Beche 1846: 221). This paper perhaps had some effect on Elie de Beaumont's change of mind; it is known to have influenced other geologists deeply (Suess 1904–1909:I, iv).

For Elie de Beaumont, this new theory was the culmination of twenty years of thought and work in which he had constantly stressed that mountains of the same age had the same or *nearly* the same direction. After a short period of conducting his geological education before the public, in which he moved quickly from six, to twelve, and then to twenty systems of mountains, he had settled on the latter figure as accurate for Europe. In his attempts to discover *why* there should be twenty different systems of mountains of known age he had tried to inscribe the intersections of their strikes on a map. In transferring these results to the surface of a globe, he discovered that all twenty of the ranges "fit" on fifteen great circles of the terrestrial sphere, the surface of which was divided by their intersections into twelve regular pentagons. This was the (in)famous *réseau pentagonal*, ("pentagonal network") which is the only part of his geology that is generally remembered—and to his discredit. For Elie de Beaumont, these fifteen lines forming the sides of the regular geometric figures had a tectonic and dynamic significance—they were what he called lines of *plus facile écrasement*—zones of weakness in the earth's crust that were the most visible manifestations of the long-term shrinking of the interior of the earth. Again, De la Beche had seen the germ of this idea in Elie de Beaumont's earlier work—that a linear zone, once dislocated, would be the place on the crust most likely to receive further dislocations.

There is no question that the *réseau pentagonal* with its increasing complexity through the years was his *idée fixe,* and a magnet for criticism. The petrologist H. Vogelsang called the system "geological-mathematical chaos" (Vogelsang 1867:88). The American orographer George L. Vose noted that the expansion of the system from the original six correlated paleontological and orogenic episodes, undertaken to avoid criticism and difficulties, had "so modified it as to destroy the only claim it ever had to attention—its simplicity" (Vose 1866:14–15). For Vose, Elie de Beaumont's emendations had destroyed the logic of the system "inasmuch as he has

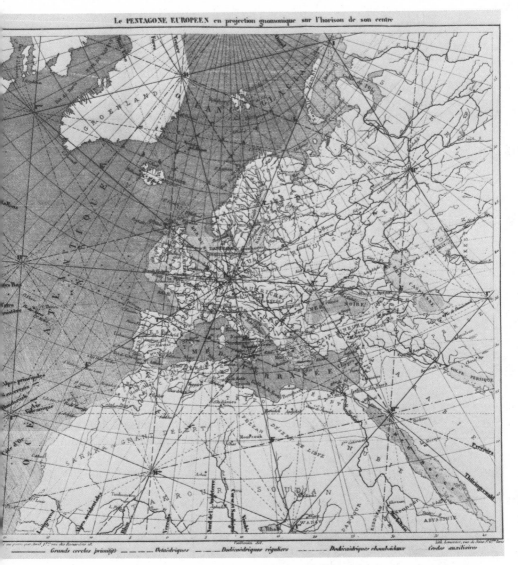

Grands cercles primitifs ⎯ ⎯ ⎯ Octaédriques ⎯ · ⎯ · ⎯ Dodécaédriques réguliers ⎯ · · ⎯ · · ⎯ Dodécaédriques rhomboïdaux ⎯⎯⎯⎯ Cercles auxiliaires

Elie de Beaumont's *réseau pentagonal*. This is the European pentagon, with the solid lines representing the "primitive great circles" and the broken lines part of the octahedral and dodecahedral networks added in later versions of the theory. Axes of principal mountain systems are found along the borders of the regular pentagon and along the altitudes from each vertex to the midpoint of the opposite side. (From Elie de Beaumont 1852: III, endpaper.)

multiplied the number of successive upheavals, and asserts that new lines of elevation sometimes take the direction of old ones, thus destroying the use of parallelism as a time boundary" (Vose 1866:15). True. But by 1850 Elie de Beaumont was no longer pursuing a simple technique in geochronology, nor was he so enamored of the theories of Cuvier and Buch, which had spurred his early theory of correlations; the criticism of Vose misses the mark in that sense at least. In Elie de Beaumont's later works on mountains there are two evident goals. The first, and the foremost in his own mind, was the validation of the mathematical inspiration of his pentagonal system, and the regularization of geology at the macroscopic level by the same avenue which Haüy had taken at the microscopic—the study of angles of intersections as determinants of important geological facts. Whereas Vose deplored the complexity of the system, Elie de Beaumont gloried in it. The lines of intersection were for him a kind of wonderful clavier on which nature had constantly played since the earth had begun to cool. At first he had thought it a tune of a few notes, but he had come to see and hear the more complex chords of the universal melody and its long-term worldwide harmony (Elie de Beaumont 1850:338).

Elie de Beaumont's goal was more prosaic—a catalogue and description of all the world's mountain ranges and their structure. In 1852 Elie de Beaumont published the longest and most elaborate version of his system and its attendant extensive descriptions of mountains. The poetry was there, but the metaphor had changed— the mountain ranges were now the capital letters of the immense manuscript of the earth, in which each total system of mountains made up a whole chapter (Elie de Beaumont 1852:3). The reference to these capital letters was only a minuscule part of the huge *Notice sur les systèmes des montagnes* of 1852, an exhaustive catalogue of the known ranges of the world, more than 1,500 pages long and the most ambitious work ever undertaken on the subject. Some years before, the editors of a planned encyclopedia had asked Elie de Beaumont to prepare an article on mountains. He had agreed but was delayed by the preparation of the volumes of explanation to accompany the Carte Géologique, which he wrote himself. When volume "M" of the encyclopedia was reached and his contribution was still not ready, the editors obligingly moved it to "S" under "Systèmes des montagnes." Eventually, Elie de Beaumont found that he had collected much more material than was necessary or appropriate for inclusion in the encyclopedia, and he arranged separate publication of a fuller version with a Paris firm. He did

not see the work as a theoretical treatise but a practical field guide
in three volumes. He expressly commissioned it to be published in
a format that would easily fit into the pockets of geologists, hoping
that the volumes might "serve occasionally as a *vade mecum* to geologists in their travels" (Elie de Beaumont 1852:ii–vi).

In its historical introduction, with his usual aversion to retraction, Elie de Beaumont repeated the material of 1829 without alteration. He spoke of the idea of relating Cuvier and Buch, discussed
the contributions of Steno and H. B. de Saussure on horizontal
deposition and later deformation of strata, and praised Humboldt
and Werner for the laws of parallelism. But since 1829 the actual
geological analysis of mountain ranges had become much more
sophisticated. Mountains were not single, linear structures, but
complex *chaînes,* each chaîne composed of a number of separate
rectilinear elements, or *chaînons.* Those chaînons did not describe
a single line nor were they necessarily parallel—they might even
intersect. However, the general trend of the complex group of
chaînons followed a given limited orientation which made up a
système des montagnes. This revision of the system away from constant parallelism, or the necessary identity of the strike of individual stratigraphical sequences with the range as a whole, is an obvious response to the criticisms of his system made by Lyell, Ami
Boué, and many others.[2] All of the mountain systems described
extended on lines of *plus facile écrasement* ("zones of weakness")—the
fifteen great circles of the réseau pentagonal. However, the principal proposition, and the point on which he would insist, was that
the mountains of the earth are not diffused at random like stars in
the sky, but follow a fixed pattern (Elie de Beaumont 1852:2). The
distinction is well chosen—his mountain systems were not the con-
coctions of the human imagination, grouping random stars into
fictitious and gratuitous constellations of objects, but represented
real and unrandom distributions of mountains across the earth's
face.

Whatever inspiration Elie de Beaumont had taken from Haüy
in establishing the pentagonal network of mountains, he tried to
avoid any association of his work with the idea that the earth itself
was a crystal—as Metherie had thought early in the century with
regard to individual mountains. Elie de Beaumont pointed out that

2. Indeed, it sounds remarkably similar to the notion developed later by James
D. Dana of monogenetic and polygenetic mountain ranges in geosynclinal belts,
which functioned like Elie de Beaumont's lines of *plus facile écrasement,* a theory
much in favor well into this century.

where crystal symmetry was fourfold, his mountain symmetry was fivefold, and therefore, he said, the imputation of ideas about crystals to him was a misconstrual (Elie de Beaumont 1852:x). Yet the very denial is coupled with a claim for priority of discovery of the geometric regularity of the dislocations of the crust, which shows how attached he was to the approach through geometrical analysis as a key to earth structure and suggests once again his hope that a direct route, like Haüy's, of establishing a series of numerical indexes might unlock the macroscopic structural plan. The index he particularly favored was the angle that any great circle made with the equator—any range lying at that angle to the equator could have its age instantaneously determined—so he vainly hoped. By the time Charles Sainte-Claire Deville reviewed the system in 1875, a year after Elie de Beaumont's death, the theory included more than 100 mountain systems, describing not just pentagons but polygons of a higher order: dodecahedra and even icosahedra. Though he struggled to save the system while grappling honestly with the outpouring of new and contradictory facts, Elie de Beaumont lived to see his own system gradually dissolving into natural history, as Werner's geognosy had slowly but perceptibly vanished back into stratigraphy.

Yet however much Elie de Beaumont remained wedded to the idea of the réseau pentagonal and to his dream of the direct application of mathematical analysis to the complexities of the mountain systems of the world, his great compilations also served to advance a number of more durable ideas. Though the account of the various mountain ranges that made up the bulk of the work ceaselessly referred to the overall scheme, its impact was, even among critics, undeniable. When James Dwight Dana opposed the geometrical speculations of the réseau pentagonal, his response was not to abandon geometric analysis but to return to another system of preferred trends—that of Humboldt (Dana 1847b:384). Prévost, identified as Elie de Beaumont's sharpest opponent, effusively praised his application of geometry to analysis of surface features. The technique of viewing mountains from the standpoint of their overall trend and dating the trends by the unconformities at their bases had become, at the hands of Elie de Beaumont, "one of the most solid bases of modern geology." Further, even though Prévost claimed priority for discovering the importance of secular contraction and of transverse crushing in the creation of mountain ranges, he accorded to Elie de Beaumont the "most solemn and definitive statement" of the import of the supposed mechanism (Prévost 1850:438–440).

In the final volume of the compendium of 1852, Elie de Beaumont elaborated further the doctrine of secular contraction, with a sophistication far in excess of some of his own followers. The cooling and shrinking of the earth was for him the ultimate cause of surface irregularity and controlled the structure, density, and disposition of irregularities across the face of the earth (Elie de Beaumont 1852:1349). While the shrinkage of the earth was slow and constant, the creation of mountain systems was not; it was the result of the instantaneous crushing of a crescent-shaped zone of the earth's crust. Yet the total radial diminution of the earth could not be estimated by measuring the original lateral extent of the packed strata of fold ranges, for during an episode of transverse crushing, the decrease of the earth's volume was quite imperceptible. This latter point was the cause of some controversy in many later versions of the theory—when it was argued by some geologists that the amount of total crushing or compacting in zones of easy fracture could not be made to conform to any reasonable estimate of the contraction of the earth. In Elie de Beaumont's version, the objection was removed, since the mountains need only express part of the total contraction of the globe.

Elie de Beaumont was more of a classical rationalist than any of his contemporaries and a good many of his predecessors. He was heir to what Cassirer called the "pan-mathematics of Descartes, the self-deception of believing that the multiplicity of the world could be overcome by mathematical ordering" (Cassirer 1950:15). The very pentagons that he saw as evidence of a greater order in nature (in a tradition that certainly goes back at least as far as Kepler's nested geometric solids in the *Mysterium Cosmographicum*) were robbed of their demonstrative compulsion by the development of non-Euclidean geometry in Elie de Beaumont's own lifetime.

The instantaneous uplift upon which he insisted was successfully contradicted by Lyell, who, as we have seen, devoted a chapter in *Principles of Geology* to its refutation (Lyell 1830–1833:III, 337ff). Nevertheless, in spite of the opposition and theoretical conservatism of Lyell and others, Elie de Beaumont was able to introduce several highly significant generalizations into nineteenth-century geological thought. First, he maintained the connection of geology with cosmology in relating individual surface structure to the total history of the earth. Second, he caused wide and compelling assent to the theory of secular contraction in spite of the opposition of the Lyellian school. Third, he suggested a theory of periodic episodes of mountain building, which emerged from a strong and stratigraphically well-founded conviction that mountain ranges could be correlated in time and by their interconnections with worldwide secular geological processes. Fourth, he put forward the attractive suggestion that fold mountains were the result of transverse compression, that the crust of the earth was composed of stronger and weaker sections, the latter being the lines of *plus facile écrasement*. Finally, he related these lines to each other across the face of the globe, convincing even his opponents that there was in them some geometrical regularity requiring explanation.

Elie de Beaumont's work shared with Lyell's great confidence in the efficacy of the erosion cycle, and like Lyell, he saw the face of the earth shaped in a continual contest of fire and water. However, in Elie de Beaumont's work there are regularities greater than in Lyell's system, which denied progress and asserted only repetition. For Elie de Beaumont the phenomena of geology demonstrated that the earth was on a journey in time: it had an origin, which might be deduced from astronomy; and it had an end, a consequence of thermodynamics. In the meantime the surface features of the earth and their relations to one another included signatures of progress along this historical course—regular and intermittent contractions of the crust that were expressed in the creation of mountain systems. Elie de Beaumont's work is a peculiar blend of many elements of thought—catastrophe, regularity, mathematical analysis, extensive field observation of complex structures; but in his hands catastrophism began to be transformed into a theory of geology in which strictly uniformitarian causes acted with periodic variation in intensity, a theory which became a worthy and durable opposition for the growing school of Lyellian uniformitarians. After 1850, the theory assumed the status of orthodoxy in France. Writing in 1911, Stanislas Meunier recalled that to 1875 "not a single article appeared, whether in stratigraphy, orography, or even geology applied to mining in which the author failed to assert, at least in the last paragraph, that the beds studied, the mountains described, that the veins in question, appeared exactly in alignment with such and such a great circle, thus confirming the views of the Master" (Meunier 1911:22). Certainly this reflects as much Elie de Beaumont's power within the French geological community as the persuasiveness of his views, and it also expressed the desire of geologists and editors to partake of the general theory of the earth as a frame for their work so that it would appear up to date. Today one finds the same phenomenon with regard to the theory of plate tectonics—no matter how conventional the subject, the author generally appends a short final paragraph tugging his forelock in the general direction of the plate margins. Between 1850 and 1875, however, the ensemble of ideas put forward by Elie de Beaumont was generally accepted not only in France but by many geologists in the rest of Europe and in North America. In spite of their oddity from the modern point of view, his theories must be recognized to have asserted a profound and lasting influence on the development of geotectonics—an influence far greater than that of Lyell.

CHAPTER FIVE

The Debate in North America, 1840–1873

WHILE the debate on the origin of mountain ranges progressed in Europe, an entire new field of exploration was being opened by North American geologists. The mountains of New England had been studied early enough that their interpretation could figure in the controversy between the Huttonians and Wernerians. But subsequent investigations in the eastern United States revealed that these mountains were only a part of a much larger system— the Appalachians, extending from New England into the Deep South. T. Mellard Reade, a British geologist who revived the work of Babbage and Herschel in the 1880s, noted that the Americans, whose rocks forced them to face "stupendous mechanical problems," had begun to devote attention to dynamics long before English geologists, who remained much more involved in the establishment of the correct stratigraphical sequence of less disturbed regions (Reade 1886:1). The principal "mechanical problem" faced by the Americans was the mechanism whereby the Appalachian system, composed of strata tens of thousands of feet in thickness, had been regularly flexed into undulatory waves, overfolded and overthrust, in a region 1,500 miles long and 150 miles wide (Vose 1866:26).

The interpretation of Appalachian structure quickly became a controversy along the same lines as the one then in progress among European geologists. Was this mountain system a product of some special process or agency that had created mountains from flatlying strata, apart from a general elevation of the North American continent? Or were the Appalachians only the remains characteristic of the more general and uniform processes advocated by the Lyellians? Both positions quickly found articulate spokesmen.

The first explication of these structures was the work of two brothers—Henry Darwin Rogers and William B. Rogers—who mapped large sections of Pennsylvania, New Jersey, Virginia, and West Virginia in the 1830s and 1840s. Their publications on the Appalachians, beginning in 1842, were recognized immediately on both sides of the Atlantic as a superlative achievement in geology (Rogers 1975:504–506).

The mapping by the Rogers brothers of the Appalachian mountain chains was the first careful and reasonably complete elucidation of the structure of an entire mountain system of the folded type. Roderick Murchison's analysis of the Urals followed in 1845, and J. Thurmann's work on the Jura, which summarized a generation of labor and more than fifty expeditions, was not completed until 1853 (Murchison 1845, Thurmann 1853/1854). As a result of the priority of the Rogers brothers, these European mountains came to be seen as "Appalachian-type" ranges. The Appalachians were characterized by regular sinuous flexures, overfolded and overthrust in a way that suggested that the entire system had experienced a lateral push from east to west. While their structure was not simple in any sense, it is not surprising that they should have been successfully interpreted before the more complex chains of Europe; the true Alps did not receive a really general interpretation until after 1875, following a century of continuous study.

Immediately notable about the work of the Rogers brothers was its clarity and accuracy: the transverse sections were drawn without vertical exaggeration; the vertical distance in feet and miles was drawn to the same scale as the horizontal (Rogers and Rogers 1843; Rogers 1858). This technical improvement in geological mapping had theoretical consequences. In 1866, George Vose compared their work with the approach of Herschel, whose schematic drawing of the Gulf of Bothnia was reproduced in the last chapter. Vose noted that if Herschel had paid attention to scale and not exaggerated his picture of the Gulf in such a way as to make it one-fourth as deep as it was wide, he would never have "fallen into the absurdity" of having sediment from the continental margins spread in even sheets across the entire floor of an ocean basin (Vose 1866:38).[1]

1. Such things move in and out of the category "absurd." Today it is recognized that turbidity currents originating in the slumping of sediment on the conti-

The Appalachians, when seen in true vertical scale, gave considerable cogency to the idea that a tangential movement was responsible for the folding. Indeed, Henry De la Beche, who was the strongest contemporary proponent in England of tangential compression as the motive force in the creation of mountain structure, had argued strenuously for more attention to true vertical scaling in geologic mapping when he inaugurated the Geological Survey of Great Britain. Certainly, symmetrical flexures over a range of 150 miles look very different when drawn as if they were 2 to 4 miles deep rather than 40. There was a further suggestion in the true-scale maps of the Rogers brothers: regular folding over such a distance might better be explained by folding *after* deposition was complete than by concurrent disturbance by subterranean vapors and exhalations as Herschel's sketches of great deeps might lead one to believe. For the Americans, the Appalachians did not appear to be merely the erosion-dissected remnants of flexed strata participating in a general continental uplift but the result of a distinct and locally expressed process of mountain making.

Armed with his maps, sections, and interpretations, Henry Darwin Rogers traveled to England in 1842, where he met Lyell, Babbage, and De la Beche, among others, and where he personally joined in the debate then in progress between De la Beche, who had supported Elie de Beaumont on tangential compression and secular contraction, and Babbage, who with Herschel and Lyell held for mountain formation by continental uplift and erosion according to the mechanism already described (Gerstner 1975). Rogers's own theory of mountain formation shows how difficult it is to make sense of mid-nineteenth-century geology by dividing geologists into uniformitarian and catastrophist camps. Rogers sided with Babbage and those other uniformitarians who held for continental uplift and constant heat within the earth, and yet his theory of mountains was among the most wildly catastrophic ever put forward. He interpreted his own superlative and painstaking fieldwork to show that a great earthquake in the Appalachian region had set up undulatory waves in the crust, as the great lava reservoirs beneath churned and sloshed about, and that these great crustal waves had been immediately frozen by injection of molten matter from beneath, producing the sinuosities of the major sections of the system. It has been recently argued that Rogers's theory

---

nental shelves can carry huge volumes of these deposits hundreds of kilometers into the abyssal plains of the oceans. After the late 1950s, Herschel's view had ceased to be "absurd."

of mountain ranges was one of America's great contributions to theoretical science before 1850.[2] Considering that his theory was only accepted by a handful of men, and then only for the few years it took James Dwight Dana to overthrow it with a far superior interpretation from the standpoint of the theory of secular contraction, the claim cannot be substantiated. Rogers's fieldwork and maps, on the other hand, were the foundation of some very fruitful theoretical ideas by other investigators, especially in the demonstration of an undeniable association of thick sequences of sediment with the formation of a mountain system.

James Hall, Jr. (1811–1898), of New York, vigorously opposed Rogers's catastrophist interpretation of the Appalachians and the view of other theorists that there had occurred some special process of mountain making along the East Coast of North America. Hall was no relation of the celebrated experimental geologist of Scotland, but he shared with him a strong preference for Hutton's views. A geologist and paleontologist, Hall was a major figure in nineteenth-century American science, noted for his outspokenness and asperity. His most important work on mountain structure was delivered as his presidential address to the American Association for the Advancement of Science in 1857 (Hall 1882).

Hall was the strictest of uniformitarians and a proponent of continental elevation in Lyellian terms. As late as 1882, when he added a few notes to his original text of 1857, he refused to entertain any dynamic notions "beyond what has been suggested by Babbage and Herschel." His work was the most painstaking sort of paleontological stratigraphy, and he deeply resented the facile schematism of those of his contemporaries who "with more or less of true science, are still so fond of making theories of their own, that they at once invent a plausible and acceptable plan which goes forth to the world with all the circumstance attending real scientific investigation" (Hall 1882:35).

Hall's position represents the natural alliance between a cautious, natural-historical empiricism and uniformitarian dogma. Like Lyell, Hall insisted that geological hypotheses should rest on a sure

2. Rogers's theory: The strata of a region are subject to excessive upward tension, the result of expansion of molecular matter and gaseous vapors below; the tension is relieved by linear fissures through which the gases escape. In their escape they produce violent pulsations in the molten matter below, which pulsations are communicated to the strata above, which are thrown into flexures and keyed to permanent forms by the intrusion of molten matter. During the oscillations the entire tract is shoved or floated in the direction of advancing waves, explaining overfolding. The claim for the importance of the theory is Gerstner's.

foundation of observation and on geological processes presently
observed to operate before invoking processes that could not be or
had not been seen. Also like Lyell, he was engaged in a two-front
war: a rearguard action against the biblical geologists and a perpetual skirmishing with overenthusiastic dynamic theorists who went
too quickly from field geology to grand theory. Hall inveighed
against the absurdity of reconciliation of geology with Scripture,
since geological knowledge expanded so quickly that "every quarter
of a century presents our science in so different a phase that a *new
reconciliation* is required" (Hall 1882:35). But he reserved his rancor
for those geologists who dipped into the work of active investigators to support one or another dynamic theory of the earth—
theories that he felt to be as unnecessary as they were speculative.
In spite of the great historical complexity of his own paleontological studies, Hall believed that "geology, if we would let alone
grand theorizing, is a simple and beautiful study, in which we see
everything evolved naturally and harmoniously, without at any time
great and sudden changes" (Hall 1882:63).

Henry Rogers's theory of the formation of the Appalachians
was characteristic of most mountain-building theories of his time—
as sudden and simultaneous deformation and uplift into a folded
range of what had been a flat sedimentary cover. Hall's position, in
1857 and after, was that such theories were gratuitous fancies that
could be dispensed with by careful analysis of the way in which the
sediments had been laid down. The history of deposition and of
the subsequent uplift and denudation of the continent as a whole
was, for Hall, the complete geologic history of any given region,
mountainous or not, and his own theory of the Appalachians was
constructed from this standpoint.

Hall began his history of the Appalachian region by tracing the
character of the rocks in the earliest Paleozoic periods, and he
noted that the third oldest group of sediments in the assemblage,
known as the Hudson River Group, had some interesting characteristics. Its greatest thickness was made up of coarse sandstones
and conglomerates and great bands of shale. The earlier (lower)
Potsdam sandstone (today the Croixian stage of the Upper Cambrian period) had been identified as a sediment deposited in shallow water, both by its paleontologic content and the presence of
ripple marks. It was at that time generally inferred that sandstone
could be deposited only in relatively shallow seas (Aubouin 1960:
138). Yet this Hudson River Group was more than 2,000 feet
thick in Canada, and according to Rogers's fieldwork (whose qual-

ity was unquestioned) the thickness of the group increased to more than 6,000 feet in Pennsylvania. The formation was consistent throughout the length of the Appalachians—beginning in the Gaspé, passing through Vermont into the proper Appalachians, and staying with them as the range declined to the south. "Along this line," wrote Hall, "which I shall term its line of trend, it presents no striking variations—nothing but what might be expected in any sedimentary group of rocks" (Hall 1882:42). It was in the passage from east to west that the interesting variations occurred. Hall traced the Hudson River Group westward along two lines perpendicular to the present mountain system—from eastern Canada to Lake Huron, through Green Bay, Wisconsin, to the Mississippi; and farther south, through Cincinnati, Ohio, to the mouth of the Ohio River (joining the Mississippi at Cairo, Illinois). Hall saw that as he traced the group farther west, the coarser sediments gradually disappeared, sandstones passing into muds and finally limestones. Moreover, not only did the character change, but the thickness decreased dramatically. A group thousands of feet thick in the region of New York from which it took its name was less than 100 feet thick in Wisconsin and Illinois and less than 50 feet thick in Iowa. This thinning to the west was not a result of erosion: all the elements of the group were still present in the west.

Hall's explanation of this gradual change was that while the present continent was still an open sea, a powerful current had brought the sedimentary materials—sand, clay, and gravel—from the northeast and distributed them along the line now marked by the Appalachian chain. If one wished to assume the presence of a former continent to the east of the United States, he said, these sediments were distributed along its coast.

> On whatever ground we view the deposit, we must admit, I conceive, that these materials have been distributed by a current, and that the direction of the current has been in the line of greatest accumulation and coarser materials; and that towards the westward, where we find less accumulation and finer materials, but still a wide ocean, the current gradually diminished until it essentially ceased, and the fine materials were slowly spread over the broad area which they occupy in their diminished thickness. [Hall 1882:42]

Hall went on to describe the similar characteristics of every succeeding Paleozoic group up to the Carboniferous period, which for him marked the final emergence of the continent in that era, except for shallow marine incursions. The entire length of time in which this great mass of sediment was deposited preserved the tendency for the great thicknesses in the east to thin and become finer to the west.

But this stratigraphy had not been at issue—the point of controversy was the means by which the Appalachian strata in their greatest thicknesses had been contorted. Here Hall turned to the agencies allowed by the uniformitarian camp. The particularly great thickness of the Onondaga Salt Group (New York State) led him to postulate a deep basin caused "in all probability by the lateral pressure on the east, which elevated the rim of the basin on that side rather abruptly while it sloped more gradually to the west. This lateral pressure was apparently caused by the gradual contraction which produced the more abrupt foldings of the crust in the region of New England and Eastern New York" (Hall 1882:48). While the immediate context might lead one to infer that he supported a limited version of the theory of secular contraction, this is not the case. He explicitly denied that he wanted to consider the folding of the crust as due to secular contraction (Hall 1882:69). The contraction referred to is Babbage's supposed contraction of deposited sediments as they were heated, which Babbage had invoked to explain both deformation and to provide room for further sedimentation. The furthest Hall would go toward any dynamic theory, it will be remembered, were the agencies propounded by Babbage and Herschel.

Any discussion of the details of dynamics was sensitive ground for the uniformitarians, and Hall in each case took the least violent path to the explanation of the appearance of the sediments. To this end he adopted the theory of chemical metamorphism of the Canadian geochemist T. Sterry Hunt, who had described the recrystallization of many rocks as a secular process carried on well below the boiling point of water. This avoided the dangerous extremes of contact metamorphism, which required great igneous outbreaks that fused or melted the sediments they touched, and dynamic metamorphism, in which rocks were recrystallized by the heat of friction and great pressure as they were squeezed. Hunt was in turn a strong supporter of Hall and reviewed Hall's theory very favorably (and extensively) in the *American Journal of Science* in 1861.

Hall also argued that the folding of the sediments had progressed gradually from the beginning of the Paleozoic era, rather than in a single episode. He saw a number of discordances in the sequence, which exponents of a more violent history for the earth might have seen as isoclinal packing, *Schuppen-struktur,* and evidence of lateral thrusts, to be angular unconformities of the Huttonian type, indicating lacunae caused by periods of erosion, uplift, and deformation at a secular pace. The "system of folding and plication therefore, which has been attributed to later violent action . . . has . . . been produced by continued action from a long anterior period" (Hall 1882:54).

The whole tendency of his argument was away from dependence on any particular mechanism of deformation, but to strengthen that argument he backed away hypothetically from the uniformitarian agencies he had invoked. Instead, he supposed (for a moment) that the folding and contortion of the strata had taken place in a single period. Given that assumption, he suggested, let us inquire "whether this has anything to do with the production of a mountain chain, or whether this system of mountain making by upheaval, overturning, etc., is a system of mountain making, or whether rhetorical eloquence has not made us believe the language without testing the facts. I must assert that it has nothing at all to do with it—that so far at least as regards our mountain chains, the principle in its application is false" (Hall 1882:54).

The principle was false because, according to Hall, folding and contortion served to *lessen* elevation, not increase it. Since he associated contortion with the contracting mechanism suggested by Hunt (and Babbage), contorted rocks had a smaller volume than the original strata. The current flowing along the borders of the eastern paleocontinent carried huge volumes of sediment for ages along what is now the axis of a mountain chain. With the accumulation of sediments, "this line was depressed time after time, as new deposits were piled upon it." Later the continent was raised to its present elevation (by a mechanism he neither discussed nor described); what elevation there was in the Appalachians was a function of the thickness of sediment deposited along the line of the current, *minus* the loss of elevation by contraction and contortion of the strata. Had the Appalachian strata not been folded, the eastern United States would be a plateau 20,000 feet above its present level: "elevation is due to deposition." The reason that there were no mountains along the Mississippi was that even though the strata were of the same kind and the continental uplift the same, the strata that far west of the main line of current were not thick enough to make the kind of plateau that could be dissected into a range of mountains. He denied that such thinned sediments might be locally uplifted by an upthrust of crystalline rocks from below:

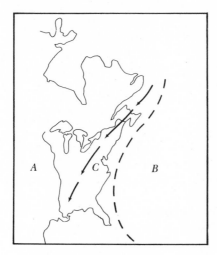

Reconstruction of James Hall's 1857 conception of the formation of the thick Appalachian sedimentary sequences. *A* is the eastern half of the present North American continent. *B* is a hypothetical paleocontinent in the Atlantic region, the source of the sediments; its western border is represented by the broken line. *C* is the strong current running parallel to the coast of the former continent, carrying the sediments to their present resting place in *A*.

"the upheaval of mountain chains, except in the sense that the elevation of a continent is upheaved, has never taken place . . . the general height of the country gives the entire elevation produced by the thickness of the beds, and there can be no more" (Hall 1882:54–56).

There is an apparent contradiction here. Hall asserted that the line of deposition was depressed time after time under the weight of sediment, yet he insisted that the entire continent was uniformly elevated from the East Coast to the Mississippi. If both these conditions were true, then the uniform reelevation of the continent could never return a portion locally depressed to its former altitude, and the great thicknesses of sediment could never have produced the plateau he postulated in the absence of folding. Yet it is clear that he did mean that former deposition was the absolute index of later elevation. Comparing the unfolded Catskills to the nearby folded Green and White Mountains, he noted that the latter were of a greater thickness than the Catskills and, had they not been folded, would have overtopped them by many thousand feet (Hall 1882:56).[3] His confusion suggests the little attention he paid on principle to dynamic details but also suggests the contradictory

3. The use of the term *contraction* can be very misleading in these discussions. In this version of Hall's ideas, it is not at first clear which contraction (secular contraction of the earth's radius or local contraction from heating and recrystallization) is meant. In the version in his *Paleontology*, on the other hand, it is as evident that he meant that the crust *subsides* and does not merely contract. See the excerpt in Mather and Mason 1939:410.

accounts of subsidence in the memoirs of Babbage and Herschel at the end of *The Ninth Bridgewater Treatise*—Babbage suggesting contraction of sediment, Herschel subsidence of the crust under weight.

Hall urged his system as an explanation not only for the existence of the Appalachians but for the outline of the continental borders of the United States to the east and west, shaped by current flow over many ages. This is, said Hall, "the simplest system of mountain making which can be proposed" (Hall 1882:55). One is inclined to agree with him. He had succeeded in suggesting that attention to only the main line of Appalachian deposition and its history, rather than to the entire sedimentary history east of the Mississippi, had given a distorted idea about the way in which the Appalachians were formed. The "trough" in which they were formed was many hundreds of miles wider than the 150 miles of the thickest and most deformed sediments. Careful fieldwork on a continental scale showed the true character of these deposits, whereas parochial focus on a "chain" had given rise to wild theories of local linear uplift.

In the version of these remarks published in 1859, Hall admitted that he had not actually followed the debate over mountain structure with much care: "I must confess myself, moreover, greatly in arrear regarding the present aspect of these and similar points of discussion." He defended his views nevertheless with the statement that he had "controverted no established fact or principle beyond that of denying local elevating forces, and the intrusion of ancient or plutonic forces beneath the lines of mountain chains, as ordinarily understood and advocated. . . . In other respects the views I have advanced are the legitimate results of observation." The extent to which he had ignored the debate is further underlined by the following suggestion.

> In bringing forward the foregoing statement of facts, and what I regard as the legitimate inferences therefrom, I have not thought it necessary to controvert the prevailing opinions relative to the elevation of mountain chains.
>
> The grand theory so beautifully and completely elaborated by E. de Beaumont, and at the present time received by a large number of geologists, may still equally apply to the exposition of the system of mountain chains: nor indeed does it appear necessary to do more than change the language of description regarding the process of elevation. If my views of accumulation and the results therefrom be correct, then the lines of mountain elevation

of de Beaumont are simply lines of original accumulation. [Hall
1859:89]

This is either charmingly naive or very crafty, and Hall's astute-
ness in other matters leads one to suspect the latter. All that need
be done to have his and Elie de Beaumont's theories conform is to
annihilate the latter and replace it with the former: without the
connection of secular cooling, transverse crushing, and local uplift,
Elie de Beaumont's grand scheme was little more than a detailed
map of the world's mountain systems.

Hall's work was successful in pointing out the need to study
regional sedimentary environments, and not merely the sediments
close to the axis of a chain, if a true picture were to be achieved.
This success notwithstanding, his work came to be criticized for the
same thing for which all uniformitarian theories were criticized—
the vague and unexplained notion of the uniform elevation of
continents. This point became the focus of a long controversy between Hall (and his supporters) and Dana, who adopted many
aspects of Hall's work but argued that Hall's theory was a "theory
of the origin of mountains, with the origin of mountains left out"
(Dana 1873a:347). That is, Hall explained the origin of mountains
with an account of how the sediments making them were laid down
but abandoned his account at the precise moment they were to be
uplifted from this trench or depression, to become *mountains*.

James Dwight Dana (1813–1895), like Hall, was a major figure
in American science in the nineteenth century. His chemico-crystal-
lographic system of mineralogy, first published in 1837, was almost
universally adopted; *Dana's System of Mineralogy*, much expanded
and revised, is still in print. His textbook of geology, *The Manual of
Geology*, was a standard textbook in the United States from 1862
until 1895, and was the bible of American geology for four decades. Occupied by his system of mineralogy in the 1830s, and
from about 1840 until 1855 concerned mostly with zoological topics, especially crustacean taxonomy, he afterward turned increasingly to the study of geology. As early as the 1840s he had begun to
write on geological topics with special attention to the origin of
mountain ranges (Rice 1915:3).

Dana was a convert to the theory of secular contraction in the
1840s and continued to elaborate his geological views within the
framework of that hypothesis for a half-century. Several times he
revised the proposed mechanism of action within the scope of the
hypothesis but remained wedded to the basic insight that the cause

of the greater part of visible geological structure was the shrinking of the earth. When George Vose published a review of mountain theories in 1866, he conceded (though he opposed contraction theory and sided with Hall) that of all the geologists who had proposed contraction as the cause of mountain uplift and folding, Dana "has applied this principle in more detail, and with far more skill, than anyone else" (Vose 1866:41).

Dana's first contributions to the topic, in 1847, were typical of the highly geometric (by later standards) theories common in mid-century, especially those theories in the tradition of Werner's followers and students. As in most such accounts, great attention was paid to historical priority and the lineage of the idea of contraction, which began for Dana with the ideas of Leibniz (Dana 1847a:96). Also like much contemporary work, it was an eclectic mix of uplift and contraction theories. In evidence, too, was the (by then) widespread disagreement with Elie de Beaumont's strict laws of correlation between the age of mountain formation and a unique range strike. Dana's principal criticism was that Elie de Beaumont was wrong "on the map": there were no straight lines from which to compose intersecting networks; mountain ranges, especially great ones, generally took a curvilinear path across the face of the earth (Dana 1847b:385). Eduard Suess later used Dana's work in a major theory of mountain structure (discussed in detail in the next chapter) in which he began from exactly this criticism of Elie de Beaumont. Dana believed that the mountains, particularly those on the rim of the Pacific, were linked through a kind of radial contraction from centers of subsidence under the ocean. His argument, based on physical deduction and speculation more than field observations, was clearly that of a scientist trying to find his way in a new field of study.

Not that Dana was inexperienced in observation—he had traveled to Hawaii and across the Pacific with the Wilkes expedition and seen volcanoes in eruption. He denied that they were safety valves, as Hutton, Lyell, Babbage, and Herschel had claimed, and disputed equally that they were the result of the contraction of the earth (Dana 1849). Dana attempted to combine geological and astronomical observations with physical theory in order to explain the general structure of the crust, having observed the contradictions of the then current theories of the European schools of geology and their inability to explain much of what he had seen in the Pacific.

He argued that the earth, like any cooling sphere, would be

depressed on the side that cooled last. Since volcanic activity had been absent from the continental platforms since the Paleozoic and was still quite evident in the ocean basins, he concluded that the oceanic crust of the earth was cooling and subsiding at a faster rate than the continents, which had long since ceased to have igneous outbreaks (Dana 1847a:94). To this speculation he attached other observations: he had seen lava pools in which the heat diminished circularly from the center and had inferred from his observations of the moon that its surface showed similar phenomena on a larger scale. From these considerations he deduced that in the earth's crust there must be such circular or elliptical areas of cooling: "The force of tension in the crust from contraction beneath is exerted to a great extent horizontally; and in a subsiding area, the direction would be nearly radial, or from the center outward" (Dana 1847b: 395). From these general, global considerations, Dana turned to the study of mountains.

In nearly all current views, mountains marked the location of fissures in the crust. This was true alike for the theories of Buch, Herschel, De la Beche, and others, regardless of the specific mechanisms that they invoked concerning the behavior of these fissures and their cause. It was known (if less generally agreed) that mountains had certain determinate orientations. This was evident in the trend of connected mountain systems like the Appalachians and the Rockies, respectively striking to the northeast and the northwest—the directions that Humboldt had suggested as characteristic for mountains in general. Dana argued that these two favored directions of strike bespoke "directions of easiest fracture in the earth's crust." In this he explicitly followed Necker de Saussure, De la Beche, and Ami Boué (Dana 1847b:390).

The probable cause of such lines of fracture in the crust was, he thought, contraction—this far at least Dana followed Elie de Beaumont, sharing his confidence in this "known dynamical cause" (Dana 1847b:395). Radiating from the centers of oceanic subsidence, lines of tension passed through the ocean floor until they met areas of lesser contraction—the continental margins. At the juncture of such areas of unequal contraction, Dana thought, there was a concentrated ensemble of fissuring, folding, swelling, and rising of the strata—in a word, mountains. Other associated phenomena would be all those observed in disturbed areas of the crust: ejection of igneous material along fissures, upheaval and uplift of the fissures, and lateral displacement as a result of the lateral pressure. Dana compared his scheme of events with the phenomena discovered by the Rogers brothers in the Appalachians, and there found his suppositions confirmed (Dana 1847a:98).

Dana's early views on this question were precisely the sort of theorizing that Hall had so furiously scorned. Dana admitted that any other scheme of uplift or subsidence could as well explain the location of the fissure lines at the margins of continents—they were *there* no matter which theory tried to account for them. But the rest of the theory, too, was shot through with vacillation. In Dana's scheme, the oceanic subsidence might be paroxysmal, slow and continuous, or even occasionally stop altogether. The upheaval of mountains was seen to take place when the crust was elastically strained, or when it broke along fault lines. The process might even reverse itself and the oceanic subsidences rise again, allowing marine transgression onto the continents. Like many another early mountain theory it explained too much (every sort of reversal and intensity of action was allowed for) and too little. The margins of the ocean, under tension, were pulled away from the continents as the centers of subsidence contracted, but Dana made them responsible for compressional folds. If the lines of easy fracture were seen to surround subsidences, why were the great island festoons on the coast of Asia convex from the mainland, as if responding to a center of action there rather than in the ocean? Here was a case of the poor old earth being forced to play tricks in order to explain mountain structure.

Dana's views did not change radically in the decade that led up to Hall's challenge of all such theories in 1857. In the meantime, the problem of Appalachian structure became more controversial and gained the attention of more and more North American geologists. The principal antagonists—James Dana, Joseph Le Conte, James Hall, T. Sterry Hunt—fought for so long over such a wide range of issues and talked so forcefully past each other that it is difficult even now to sort out the alliances and decide who supported whom about what. Generally, Hunt supported Hall, but not on every issue. Occasionally, as in the debate between Robert Mallet and Joseph Le Conte, squabbling over priority of discovery, or more absurdly, priority of suggestion of as yet unsubstantiated hypotheses, made opponents of men with substantially the same views (Mallet 1873:302; Le Conte 1873:156).

By 1866 the basic outlines of the debate were becoming so blurred that a general clarification of the issues was necessary. Such a clarification came in a monograph devoted to the debate on Appalachian structure and its implications for mountain formation in

general: George L. Vose's *Orographic Geology, or the Origin and Structure of Mountains: A Review* (1866). Vose was an opponent of the
theory of the secular contraction of the earth as an agency important in the creation of mountain ranges and was more sympathetic
to Hall than to anyone else. In this work the most important service
Vose performed was pointing out that each of the major theories
contained notable absurdities, either physical or geological, not one
being tenable in its original form; this forced a series of public
reconsiderations by all participants.

By the early 1870s (once the layers of priority fights and minor
squabbles had been removed), the contributions of the various disputants began to reveal a common ground, more or less along the
lines laid down by Vose. It was agreed that there was a real relationship between deep accumulation of sediments and the subsequent creation of mountain ranges. Whether ranges appeared
according to Hall's scheme or that of various theorists who held for
orogenic episodes in the evolution of the crust was the issue still to
be decided. All, with the possible exception of Hall, allowed some
form of contraction of the earth as a whole as at least a partial
cause of the evolution of the long, linear, sediment-filled depressions of the crust, of which the Appalachians were an example. It
was agreed that alteration of sediment at depth by heat and pressure was a crucial process in the evolutionary sequence. Indeed, the
whole idea of a mountain range as the product of a unique sedimentary environment was generally agreed to be a real phenomenon and to require an explanation. This in itself was something of
a victory for the orogenic theorists. With the abstention of Hall,
few of the disputants would agree any longer with Lyell that continental uplift *alone* could explain the character of the Appalachians. On the other hand, it was equally clear that a simple law of
radial contraction, combined with unexplained "lines of easy fracture" in the crust, was not an explanation in any way superior to
that of continental uplift.

Dana took the opportunity, in this new round of argument, to
make a survey of the entire question, historically and scientifically.
In a series of long papers written in 1873, he elaborated a theory of
mountain ranges and of the origin of continents and oceans that
addressed each of the major questions—the mechanism of subsidence along the chain, the various proposals for elevation (local and
continental), and the theories of deformation and metamorphism
of rock that might accompany them. He reviewed and criticized
Hall's arguments once more and corrected and expanded his own

thoughts on the subject, which went back to the papers of 1847 (Dana 1873a; 1873b). He also sought to reconcile and where possible eliminate the variant theories (Hunt, Le Conte, and others) that had created a confused middle ground between the original extremes of interpretation. "The remark," he began, "that in Professor Hall's theory of the origin of mountains the elevation of mountains is left out . . . was made by me . . . and not without due consideration" (Dana 1873a:347).

Dana's "due consideration" referred to Hall's theory of uplift and subsidence. Dana accepted the shallow-water origin of the Appalachian strata and the constant subsidence that this implied along a line parallel to the strike of the present mountain system, created out of the great thicknesses of sediment. But the mechanism of subsidence, which Hall had adopted from the argument of Herschel, Dana took to be a sheer *physical* impossibility. It had been argued that the weight of the sediment would depress the crust and, with some combination of contraction of the materials, make way for a new series of deposits above. However, argued Dana, "no reason is given why sediments under water should have so immense gravitating power, when the crystalline rocks of the Adirondacks, piled to a height of thousands of feet *above* the water, had a firm footing close along side of the subsiding region" (Dana 1873a: 349). Actually, Herschel had given a reason—the deposits in the subsiding region had upset a balance and started a continuous rise parallel to any region of subsidence as the fluid matter migrated laterally beneath. However, Dana argued, in a shallow ocean (dictated by the nature of the sediments), long before the sedimentary load could begin to depress the sea floor, the ocean would fill up and bring sedimentation to a halt. The weight was not great enough for the suggested process to get under way, even if such a process might conceivably operate in other circumstances. The subsidence was *there,* but the explanation given by Hall was an inversion of causality: "But while the weight of the accumulating sediments will not cause subsidence, a slow subsidence of a continental region has often been the occasion for a thick accumulation of sediments" (Dana 1873a:349). Some process outside the sedimentary cycle had to be responsible for the depression of the crust; Hall's simple mechanism for subsidence of the region was not possible on physical grounds alone. Consequently, Hall's picture of elevation, folding, and metamorphism, and Hall's insistence that the height of mountains equaled only the thickness of deposits, also came into question.

Hall had ridiculed those theorists who made ad hoc adjustments in their theory of the elasticity of materials in order to explain the various stages of mountain building. Yet in his own theory, countered Dana, Hall "accounts for plication by simple subsidence, but supposes the continents to get up high in some way—the way not considered—without other plications, or any local uplifts, and on a crust so flexible that it will sink a foot for every foot of sediment added to the surface" (Dana 1873a:350). Here it was Hall making the "poor old earth" play tricks. Hall's theory joined a physically impossible account of subsidence to an unexplained continental uplift, which, in addition, contained the idea that strata were flexible and deformable on the way down but inflexible and strong on the way back up. Moreover, the account of uplift ignored the manifold instances of unequal uplift of different portions of the same formations: "the world abounds in cases in which part of the sea-border deposits of a period are but little away from the old level, while other portions are many thousands of feet above it. . . . No principle has been found to explain such facts except that of local elevation" (Dana 1873a:350).

Dana's account was not entirely fair. Hall, with Hunt, had favored Babbage's idea of the contraction and alteration of sediments while they were still in a state of weakness: during the period of elevation they would already have become fused into hard rock and thus be able to resist deformation. Further, on the question of subsidence, while suggesting that he accorded with Herschel's dynamic ideas, Hall had favored a mechanism by which actual decrease in volume of the sediments created most of the room for new deposits—as in Hunt's theory of chemical alteration. As Vose had suggested in 1866, the crux of the debate was the theory of metamorphism.

Dana turned from his consideration of Hall's theory of uplift to a consideration of the supposed process, which might allow for thick accumulation of shallow-water sediments, and argued against any contraction in the volume of the sediments or the crystalline rock beneath. Again, he stressed *physical* possibility. "No experiments on rocks that I have met with have authorized the assumption that the ordinary law of expansion from heat would have been set aside" (Dana 1873b:427). The sediments, if anything, should have increased in volume when heated. On the question of superheating, metamorphism, and recrystallization of the basement rock—an argument with more physical cogency from the modern standpoint—(Holmes 1965:1124), Dana turned to geological evidence for a refutation: the basement rock underlay the earliest sediments unconformably, and any recrystallization there could not be a product of later depositions. If a period of erosion had intervened between the lowest existing formations and the basement crystalline rocks, their alteration could not very well be the result of a process associated with the current Appalachian assemblage.

Dana argued that lateral pressure (or lateral force, tension, horizontal force, tangential force, tangential thrust—he was not concerned with the semantic particulars) from deeply subsiding ocean basins pressed the ocean floor like the butt of an arch against the continents, rendering areas within 300 to 1,000 miles of the coast subject to profound oscillations in level, producing accumulation of sediment, plication, fracture, and uplift, culminating in elevation of the highest mountain chains, volcanism, and metamorphism. The inference that pressure came from the side, and especially from the oceanic side, was based on the evidence that the greatest folding and metamorphism in mountains was on the side that faced the ocean. Further, in explaining the relation of oceanic subsidence and border uplift, he noted that, as would be expected, the greatest mountains face the deepest oceans, the sources of greatest lateral pressure. This same ensemble of phenomena had been explained by Herschel, Hall, T. Sterry Hunt, Joseph Le Conte and others, but given the weaknesses of their ideas of subsidence under load Dana argued: "In the present state of science, then, no adequate cause of subsidence has been suggested apart from the old one of lateral pressure in the contracting material of the globe" (Dana 1873b:427).

This argument on the ultimate source of phenomena was not apt to convince those, like Hall, who detested the very search for such ultimates: once it could be agreed that subsidence occurred and that deposition and attendant metamorphism might somehow be involved in deformation, the Lyellian side abandoned the field as leading to unnecessary speculation. Nevertheless, the question of uplift remained. Hall had held to the idea of uniform continental uplift in true Lyellian fashion. Hunt had supported the idea but was tottering toward some notion of local dynamics. Le Conte had argued for continental and local uplift, but the latter only as an artifact of tangential compression. In this confused area, it was Dana's turn to deplore speculation and dogmatic adherence to one or another agency. There had been too much attention to the idea of *single* ranges, Dana argued, and to the development of a single explanation for them. In the development of a complex mountain system, composed of many ranges, there was ample room for the expression of every sort of mechanism. Why should it always be the same in every component range?

Dana's argument for a complex system with many determinate stages of development was conciliatory insofar as it did not seek to refute the possibility of any particular explanation of structure, locally expressed. It reflects Hall's suggestion that too great an attention to any part of a widespread and complex environment would distort the meaning of the whole. The idea of a system with many component ranges also suggests Elie de Beaumont's notion of *chaînons, chaînes,* and *systèmes* as successively greater units describing the structural history of a mountainous region. Systems of mountains like the Rockies and Appalachians were, for Dana, "polygenetic assemblages" of "monogenetic ranges," each monogenetic range comprising a separate structural episode. A monogenetic range began with a downfold in the contracting crust, "carried forward at first through a long-continued subsidence, a *geosynclinal*" (Dana 1873b:430). This is the *first* appearance of this term in the history of geology. The notion of synclinal and anticlinal folds was a familiar geometric description of warp of the crust, but in Dana's usage, heralded by the prefix *geo-*, there was a special historical sense and an intimation of scale. The geosynclinal downfold was an extensive phenomenon and the expression of a long-term process working in the crust. The geosynclinal was not a smooth sinusoidal surface but a large, depressed trough that contained, or might contain, many component synclinals and anticlinals as smaller depressions and swells within the regional downfold. Dana offered the Alleghenies and the Green Mountains as examples of monogenetic ranges born in a geosynclinal. The slow subsidence of the crust was accompanied by sedimentary in-filling; continued subsidence would eventually produce a climax of pressure that would introduce an epoch (of unspecified duration) of crushing and plication and ending in a "catastrophe" of solidification with incidental metamorphism and igneous activity. At any point in the process, folding and block faulting might produce great elevations, without, however, reversing the long-term geosynclinal downwarp.

Dana allowed that Le Conte had been right to assert that uplift was due to lateral crushing and that Hall was also correct, once the causality of his sequence was reversed. However, both their theories applied only to monogenetic ranges and did not explain the total dynamic history of a region. Dana argued that each monogenetic episode, from the beginning of subsidence to the end of the period of folding, was a step in the growth of the continental mass as a whole: "The process is one that produces final stability in the mass and its annexation to the more stable part of the continent" (Dana 1873b:431). Some sort of marginal growth of continents had, again, been suggested by Herschel some years before, but Dana developed the idea in a direction that overcame the difficulty in Herschel's theory, that there was no way to reverse the process of subsidence.

For Dana, the history of any range did not end with the closing of the particular monogenetic stage that gave it existence. Once annexed to the continent, the monogenetic range took part in the subsequent movements on a grander scale. The distinction between the monogenetic stage and the subsequent history of the region was to Dana the fundamental distinction in orography, all too often ignored in the past. Most mountain systems were polygenetic composites, the result of successive geosynclinal subsidences, with independent episodes of folding and crushing and local elevation. However, the great altitudes of ranges like the Rockies were not due to these local processes but to the elevation of the entire polygenetic mass by a great *geanticlinal* movement—a true elevation of a wide area of the crust.

Dana's theory of 1847 had depended on "arching pressure," a notion that other geologists had found difficult to accept, particularly Vose, who had argued that centers of oceanic subsidence should produce tensional rifting and not compressional deformation at the continental borders. To get around this difficulty and the difficulty of elevating the polygenetic mass, Dana modified the idea of arching pressure. He argued that as the continents had grown at the margins through the creation of monogenetic ranges, the crust had become stiffer and more resistant to the force of contraction, which was successively less able to produce the folds it created in an earlier and thinner crust. In the present "only feeble flexures of vast span are now possible" (Dana 1873b:433). Dana's new model of the shrinking crust had the continental crust writhing slowly "between the arches of the ocean"; through time the areas that were flexed became larger and larger, and lower in angle of deflection, though on a continental scale there was ample elevation for high relief. In short, the development of the earth was moving with an ever slower cadence and in the present had already slowed so much that there were no more geosynclinals undergoing transformation into monogenetic ranges (Dana 1873b:432).

Such broad shallow folds of great span could explain not only the reversal of direction in geosynclinal complexes (polygenetic masses) to obtain real elevation, but, if imagined on the floor of the ocean, a great geanticlinal could explain the existence of oceanic transgressions of the continents, which most contraction theorists had explained by catastrophe. In Dana's scheme, as the ocean floor geanticlinal flexed upward the oceans overran the continental margins—as the floor receded (and the continents flexed up) the oceans returned to their beds.

This theory proved extremely popular through the next twenty-five years (till 1900). Its most attractive feature was the way in which it made room for previously antagonistic theories as aspects of a still greater process. The theory of broad geanticlinal folds, while founded on the theory of contraction, amounted to an acceptance of general, if not entirely uniform, continental elevation as a source of relief (though not of structure). The separation of folding and uplift into monogenetic and polygenetic stages of development was another placating element for the proponents of general uplift, who had always urged that folding was an artifact of the epoch of subsidence. The theory avoided the tangles of metamorphic processes by reverting to the ground of lateral pressure in explaining local subsidences of the geosynclinal type. It avoided the overschematism of theories of catastrophic uplift or "once-only" movements of any kind by explaining present appearances as the result of periodic episodes of intense folding and crushing ending in a general, secular, continental uplift.

Moreover, a single, persistent dynamic cause—secular contraction—was held responsible in its various modes of action for sedimentation, folding, volcanism, metamorphism, creation of mountains, continental uplift, oceanic depression, and the alternating transgression and regression of the seas. Not only was this theoretical unity pleasing to many who wished a full account of the history of the crust according to a single mechanism, but it allowed those who wished to demur to do so—anyone who denied the overall theory could still be encompassed by it. A geologist who wanted to stay within the bounds established by Hall could be interpreted as confining himself to an elucidation of a single stage—the origin of monogenetic ranges, for instance, or the effects of the general uplift.

Dana's theory made room for a much wider synthesis of data than had heretofore been possible, by blunting the longstanding antagonism between the theory of general uplift and the theory of local creation of mountain ranges, seeing both as aspects of a single, global process, and urging absolute primacy for neither. Yet his work tends overall to be markedly nonuniformitarian; he promoted a history of the crust characterized by ever-diminishing activity as the earth's contraction lost its deforming power and was resolved more and more into broad general alternations of uplift and subsidence occurring only on a continental scale. The theory urged a fluctuation in the intensity of action through time, with folding and crushing having reached gigantic proportions at intervals in the past. Further, it argued a distribution of deformation across the surface of the globe that was concentrated at the continental margins in the geosynclinal downfolds. Moreover, the growth of continents by accretion of parallel monogenetic ranges gave a new explanation for the laws of strike of chains—their identity as zones of weakness and easy fracture was preserved without the single catastrophic action represented by earlier theories, and their constant direction could be referred to the shape of the continental core around which they had grown.

Dana's work advanced all of the fundamental points of the theory of Elie de Beaumont—the connection of the structure of individual surface features to the total history of a contracting globe, the theory of periodic mountain building, the primacy of transverse compression in producing folding, the division of the crust into stronger and weaker portions, the significance of the "grand outline features" of the globe in any understanding of its history and mechanism of activity. It combined close study of complex structures with inferences from cosmology and sought a unity of theory at a level that Lyellian geology would not countenance. Dana's work soon became a part of the continuing struggle in Europe over the nature and meaning of mountain ranges.

CHAPTER SIX

The Problem of the Alps and a Solution: Eduard Suess, 1875

In the late eighteenth and early nineteenth centuries, the debate between the Huttonian and Wernerian geologists ended in a stand-off that was both theoretical and practical: elements of both approaches survived. When theoretical debate on the "principles of geology" began again about 1830, the outcome was also divided. The "diluvial geologists" in Great Britain ceased their attempt to correlate any specific set of phenomena with the biblical deluge, while a larger and looser alliance that included many former Wernerians continued to support various nonuniformitarian conceptions of geological change and to oppose the views of Lyell and his supporters. This division on the rate and nature of geological change was most striking when the origin of mountain ranges came to be considered, and a large proportion of continental geologists, with allies in Great Britain, held theories of mountain origin that invoked periodic episodes of intense activity on a uniformitarian background to explain the large features of the earth's crust. Many geologists repudiated catastrophism and catastrophes but argued nonetheless for a long-term tendency of the earth's crust to collapse and shrink in a regular way, expressed most notably by chains of mountains.

Lyell's supporters, particularly Babbage and Herschel, tried to overcome the opposition by developing a plausible dynamic account of the origin of mountains within the confines of regional or continental elevation and the specifics of the erosion cycle. They were supported by other geologists like James Hall, who, out of empirical caution or theoretical diffidence, tried to keep geology within the limits of natural history alone.

The opposition of these uniformitarian and progressionist groups was expressed in the United States in the debate over the structure of the Appalachian region. James Hall interpreted the geology of the eastern United States in strict Lyellian terms and opposed the early catastrophic theory of Henry Rogers. James Dana, whose first theories on the structure of the Appalachians were speculative, geometrical, and cataclysmic, later developed a theory that provided a common ground for geologists of different persuasions: he suggested that the general theory of continental uplift and subsidence could, and should, include a special theory of the origin of mountains in geosynclinal depressions at the continental borders. Dana repudiated the exaggerated geometrical formalism of Elie de Beaumont and embraced the geological time scale of Lyell and Hutton for *every* process in the development of the earth's crust. However, Dana also stressed the importance of the earth's long-term contraction and cooling for the uplift of continents and the creation of mountains and voiced the distinctly nonuniformitarian conclusion that the cadence of geological change was slowing; both ideas helped move the trend of geological theory further away from strict Lyellianism and toward an eclectic and compromising middle ground.

Dana's theory was based primarily on the geologic structure of the North American continent and its relation to bordering oceans, and thus his geosynclinal theory, however attractive, had at first little obvious relation to European geology and the tectonic problems faced by European geologists. Whereas the great mountain systems of the New World ran parallel to the edges of the continents, most of the mountains of Europe were transverse, and well within the continental borders. Foremost among the transverse mountain systems of Europe were the Alps, whose complex structure had frustrated a century of attempts to correlate it with general theories of every kind. The assemblage of mountain chains made up of the Western Alps, the Eastern Alps, and the Carpathians (and, perhaps, the Pyrenees, the Jura, and the Dinaric Alps) describes a broad curve from east to west across the continent, which suggested structural continuity to those geologists who put any faith in the trend lines of a mountain system. However, repeated studies of the stratigraphical succession in the Alpine ranges failed to disclose any notable continuity along the strike of the system—the Western and Eastern Alps appeared to be made up of different sedimentary formations. From north to south, the echelons of component ranges had an equally puzzling lack of coherence—chain upon chain with distinct petrological characters, different sedimentary facies, and structural discontinuities. After seventeen traverses of the Alps between 1775 and 1796, Horace Bénédict de Saussure, the first great Alpine geologist, abandoned them as a "hopeless jumble."

The complexity of Alpine formations and structures gave room for different interpretations of the origin of the system. The limestone pillars of the High Dolomites seemed, to Leopold von Buch, to be evidence of a catastrophic elevation. Bernhard Studer's studies of the granite massifs of Mont Blanc and the Aar led him to develop a theory of the Alpine system based on plutonic upwelling along the axis of the major component ranges, with consequent bilateral folding and dislocation of the strata on their flanks. Geologists who advocated general continental uplift or dislocation by earthquake or transverse crushing of a subsiding crustal segment or various combinations of these mechanisms were able to advance their theories by judicious selection from the abundance of Alpine forms and structures. Many geologists who remained aloof from theoretical debate nevertheless made pilgrimages to the Alps to study more limited problems in stratigraphy, petrology, and paleontology and applied the abundant phenomena of the Alps to the development of these increasingly specialized fields.

Still, the overall interpretation of the Alps remained a tantalizing goal. Long before the riddles of the Alps were solved, it was agreed that when their complications were finally reduced to order and they were understood as a unit, the solution would have momentous implications for geological theory in general. From the time of Saussure, an ever-growing faction of geologists echoed his judgment that "it is above all the study of mountains which will accelerate progress in the theory of the earth."[1] By the middle of the nineteenth century no general theory of the earth that could not give an account of the origin of folded mountains of the Alpine type could expect to receive serious consideration, particularly in continental Europe, where geological thought came to be dominated by the problem of the Alps as much as the continent was dominated by the mountains themselves. The origin of the Alps had been a point of contention as early as the time of Hutton, who

1. This statement by Saussure, from the *Discours préliminaire* of *Voyages dans les Alpes,* served as the epigraph of many geological books in the nineteenth century. Here it is taken from the title page of Vose 1866.

drew extensively on Saussure's observations to elaborate his *Theory of the Earth*. Buch and Humboldt spent a great deal of effort on the geology of the Alps, and much of Elie de Beaumont's mountain theory was drawn from his study of the Alps as a classical example of the fold-mountain structures that made up the réseau pentagonal. Lyell, Sedgwick, Murchison, and De la Beche traveled to the Alps to experience firsthand their structural complexities. By the middle of the century they had become an essential part of the education of any student wishing to master geology.

One such aspirant was the Austrian geologist Eduard Suess (1831–1914), who began his geological apprenticeship at the age of nineteen.[2] Suess had joined in the revolutionary ferment of 1848 and had been imprisoned. Released from prison by the intercession of the head of the Imperial Geological Survey, Wilhelm von Haidinger, Suess was unable to obtain admission to the Technische Hochschule because of his political record and began his geological career as a clerk in the paleontological section of the Imperial Geological Museum. In 1851, he was commissioned to make a section across the Dachstein region of the Eastern Alps.[3]

Suess's Dachstein section was the first of many investigations of the Alps during a career in which he benefited from the experience of many great teachers. Foremost among these were the Swiss geologists Bernhard Studer and Arnold Escher von der Linth, whom he met in 1854. Studer was professor of geology at Berne from 1823 until 1872 and a pioneer in the detailed mapping of sections in the Alps, beginning with his traverse from Lake Thun to Lake

2. C. E. Wegmann (1976:143–148) is the source of this biographical information. Suess's memoirs (1916) are mostly concerned with his important political career and barely mention his later geological work. Another source of information about Suess is George Sarton 1919: esp. 390–392. See also Emil Tietze 1916:330–556.

3. Suess 1904–1909:I, iii. This is the authorized English translation. The Imperial Geological Survey had just been inaugurated in 1849, after considerable agitation by Wilhelm von Haidinger (1795–1871), Suess's mentor. The mapping of "sections" was a mainstay of early tectonic work, and was modeled after the work of Saussure and Studer in the Swiss Alps. After preliminary reconnaissance suggested areas for further study the geologist walked (climbed, scrambled) along a pre-planned route across a chain of mountains, and mapped the relations of the exposed strata and their thickness and extent. The results of a number of such sections at intervals were assembled into a picture of the entire chain. When the results were collated with geodetic surveys of elevations and the strike of the chain and its component branches, a geologist might attempt a three-dimensional reconstruction of the chain and think about its genesis and major lines of movement. Further analysis, often an effort of great imagination, was palinspastic mapping, in which the geologist mentally unfolded the strata and tried to determine their extent before dislocation and distortion.

Geneva, published as *Geologie der westlichen Schweizeralpen* in 1834 (de Margerie 1946:cii). Escher, the son of Hans Konrad Escher, was a second-generation Alpine geologist whose work on large-scale overthrusts revolutionized ideas about the extent of lateral movement in the Alps. Studer was a proponent of the theory of elevation craters in the Alps; Escher steadfastly refrained from any discussion of theoretical ideas whatever (Suess 1904–1909:I, iv). Escher and Studer worked hand in hand for many years on the first comprehensive geological map of Switzerland, published in 1853; Studer alone wrote the thousand-page *Geologie der Schweiz* to accompany the map. Suess considered both men as mentors and friends and remained closely associated with them throughout their lives.

Suess's ideas matured slowly, and it was not until 1875, after a quarter of a century of study, that he undertook a synopsis and critique of Alpine theories in a deceptively thin volume entitled *Die Entstehung der Alpen.*[4] With the critique and a concise overview of the rapidly growing international literature on the structure of Alpine-type mountains around the world, came a new theory of the Alps deduced by Suess from his own fieldwork as well as from the investigations of others. The work was influential all out of proportion to its size, and some years later Pierre Termier recalled that his teacher, Marcel Bertrand (one of the originators of the nappe theory of the Alps and himself a student of Suess), had decided to become an Alpine geologist only after reading *Die Entstehung der Alpen.* Bertrand, who had managed to sleep through Elie de Beaumont's lectures at the Ecole des Mines, was convinced by Suess's work that the Alps indeed contained the key to all the major problems of geology; not even the more famous work, *Das Antlitz der Erde,* on which Suess's reputation rests, made a greater impression (Termier 1908:163–164).

Die Entstehung der Alpen was the theoretical basis of much of Suess's later work. In it, he tried to clear the ground of the various hypotheses concerning Alpine structure then in vogue and start afresh. He shared the conviction of his contemporaries that the Alps were formed in the same way as the other great mountain chains of the earth and that the results of the study of Alpine formation applied as well to relief in general, apart from the destructive and constructive activity of air, water, and ice (Suess 1875:

4. A very brief selection from this book is available in Mather and Mason 1939:503–506. All citations here are from the original and are my own translations.

2). In the context of the times the work might as well have been titled "The Formation of Mountains"—the import was the same. Suess saw the Alps as the "type range" for folded mountains generally and believed that the understanding of their structure would solve all the major problems of dynamic geology.[5]

Knowledge of the variety of Alpine structures had increased apace with the progressive sophistication of geologic mapping in the decades since Suess had begun his own work in the Alps in 1849. However, he was struck by the fact that in spite of the increased volume of work on these structures and the improvement in its quality, geologists were no nearer an agreement on the nature of the force that had produced the complicated folding of the Alpine strata. While each new study reinforced the opinion that the deformation was much greater than earlier investigators had imagined, the widest possible variety of opinion still remained as to the cause of the great deformations and dislocations. The prevailing opinion, which had originated in and was particularly strong in Germany, was that the protrusion of some sort of mass, whether rigid, semirigid, or fluid, along the axis of the range had forced the overlying strata apart. While many versions had been proposed, this theory of mountain building still found its clearest expression in the work of its originator, Leopold von Buch.

Opposed to the loose cluster of theories favoring some sort of vertical uplift were the opinions of those geologists who, like Elie de Beaumont, had finally decided that some sort of transverse or horizontal pressure had first crushed the strata together and then uplifted them into mountains. Suess was critical of both positions. The theory of plutonic uplift, he argued, was not a coherent ex-

5. This statement somewhat oversimplifies the picture, as will later be seen. Following G. K. Gilbert, Suess distinguished fold mountains (*Faltengebirge*) from fault mountains (*Kettengebirge*). It was recognized that some mountains were topographical expressions of the relative vertical displacement of blocks of crust along fault scarps. Beyond this fundamental distinction of mountain types, Suess subdivided fold mountains into two groups. One group, which Suess called "parmas," were ranges like the Urals and the Jura in which the folds were more or less regular and sinuous, without overfolding and overthrusting on a large scale. The other group were the authentically Alpine mountains in which the folds took a much more complex form and were subject to complex dislocations and deformations. That the Alps could be taken as the key to all mountain ranges was a consequence of the occurrence in the Alpine system of all the above forms. The complexly folded Alps abut both the symmetrically folded Jura range to the northwest (considered a part of the system by Suess) and also a number of crystalline massifs faulted into *Kettengebirge* to the north and northeast. An explication of the Alps, as in *Die Entstehung der Alpen*, justified the system's general importance in the simultaneous understanding of the various forms in their complex occurrence.

planation but a loose conglomeration of a variety of opinions about a variety of possible processes, united only by a general predisposition to believe that the generative force came from below and was expressed vertically (Suess 1875:2). But the contrary theory of transverse crushing was also flawed; Suess quoted with approval the opinions of Constant Prévost, who had argued that one should speak only of displacement and not uplift, since from appearances either uplift of a region or a peripheral subsidence around it was possible (Prévost 1840:186).[6]

Although he thought that both doctrines were unsatisfactory, Suess noted that the greater part of current work (1875) on the subject had leaned toward a new version of the idea of transverse crushing; now, no absolute uplift was implied or required to produce mountains. Rather, the mechanism proposed was the cooling and contraction of the earth and the relative subsidence of adjacent segments of the crust. Suess referred here to a group of opinions as loose as that of the uplift theorists—he cited the recent work of Dana and Le Conte and of the Alpine geologist Charles Lory as examples (Suess 1875:4–7).[7] Suess was prepared to agree that the weight of evidence was on the side of the theory of transverse motion and subsidence, rather than vertical motion and uplift. He therefore began his work with detailed criticism of the uplift theory, addressing it in its most recent and sophisticated version: that of his friend and teacher Studer.

Studer had held (like many geologists before him) that the Alps were composed of three roughly linear zones running parallel along an east-west axis. A middle zone, surrounding isolated plutonic masses, was flanked to the north and south by linear, highly folded zones (*Nebenzonen*) of sedimentary rocks. Studer's view was that a

6. Prévost's claim to priority in the clear formulation of the theory of contraction as a tectonic "motor" was acknowledged by many geologists, including Dana. Prévost was one of the first to offer the analogy of the earth to a piece of drying, wrinkling fruit, an analogy taken so seriously that it is possible to find in geology texts printed in this century photographs of drying apples.

7. Suess had familiarized himself with the debate over Appalachian structure chronicled in the preceding chapter and had noticed that Dana's concept of a geosynclinal trough bordered by a geanticlinal upland as a source of sediment brought Dana's later theory closer to the work of Herschel. Suess was critical of the idea that the sinking and softening of an extended area of sea bottom could form the huge extensive regularities of Alpine ranges and particularly the overthrusts observed in them. In his mind "a causal relationship between geosynclines and mountain formation can scarcely be admitted" (1875:97). Later, Suess would become increasingly critical of the attribution of dynamical significance to borderland geosynclines and especially of the reinterpretation of his Alpine foredeeps as geosynclines. See Suess 1904–1909:IV, 627.

powerful uplift along the axis of the middle zone had exerted a deforming pressure bilaterally against the north and south. This mechanism of action was exhibited, Studer thought, near the Finsteraarhorn, one of the great granitic mountains of the Swiss Alps. There, granitic rocks appeared to be thrust up through fossil-bearing strata of the Mesozoic era. To geologists of Studer's day this seemed to support the idea that crystalline rocks like granite could be emplaced in the same way as lavas, which contradicted the neptunist position that granite was a primeval rock. The piercing of the later sediments showed that plutonic rocks were active geological agents throughout the history of the earth (Suess 1875:9).

Suess saw the entire argument to be as dated as the context in which it was originally raised. In Suess's view the idea of a "central eruptive" force was powerless to explain dislocations scores of miles north and south of the site of uplift. Even if the plutonic masses of the middle zone were eruptive, as they seemed to Studer, this was of little consequence—the plutonic rocks had subsequently been identified as older than the nearby sedimentary sequences they were supposed to have dislocated. Finally, the entire argument was based on what Suess considered a specious analogy between volcanic and plutonic processes that dated from a time when Vesuvius was thought to be a crater of elevation; it reflected Buch's notion of the virtual identity of all igneous activity, whether volcanic or plutonic. This equation of volcanic and plutonic activity had lost its argumentative force in supporting uplift as a uniform dynamic process when uplifted and deformed lava flows were discovered conformably bedded with sedimentary strata; these deformed lava flows could not be the cause of their own folding. With what was currently known of the action of volcanoes, Suess continued, it was impossible to assert any longer than any known group of rocks could, by some amazing manifestation of strength, lift an area of many miles, split it to the north and south, and fold it. Far from such universal active power, plutonic and volcanic rocks were seen to occur passively folded, the recipients rather than the initiators of deformation. In any case, the absurd contention that rocks of Permian age could actively uplift rocks deposited in the Tertiary could no longer be maintained (Suess 1875:10–12).

Suess attacked the uplift hypothesis on its own ground and impeached the evidence that the best of the uplift theorists, Studer, had chosen to support his views. He showed that the uplift hypothesis was overgeneralized from limited instances, that it was based on outdated concepts of petrology and interpreted on the basis of an incorrect stratigraphy, and that it invoked deforming forces greater by order of magnitude than any observed in eruptions.

This analysis of the uplift theory was a preliminary; in the main line of his argument against uplift Suess followed the tradition of trend-line analysis pioneered by Buch, Humboldt, and Elie de Beaumont. Whatever arguments might be made by close analysis of the petrology and stratigraphic succession of isolated locales, a stronger argument could be made from the evidence of the major units of the Alpine system taken as a whole. For Suess, the most comprehensive reason for rejecting the theory of uplift in the Alps by "central massives" or central eruptives pushing along a medial axis was the contrast between the irregular, broken outcrops of the supposed centers of elevation (*Erhebungscentra*) and the uninterrupted regular strike of the folds in the outer zones of the range, the northern and southern Nebenzonen. An overview of the system revealed to Suess a kind of movement in the Alps that was greater and more uniform than had been imagined: "A look at the Mont Blanc and Finsteraarhorn massifs, and from there to the great anticlinal that passes in a wide curve through the Molasse from Salève to Baiern, indicates clearly enough that the general overthrust onto younger beds lying at the north foot of the Alps could not result from the elevation or eruption of isolated central masses but only from a general movement of the entire mountain system" (Suess 1875:13).

Suess's startling hypothesis of the lateral movement of an entire mountain system was opposed equally to the uplift theory and to the current theory of transverse crushing; it had been asserted by the advocates of lateral compression that the mountains were thrust up and folded by symmetrical pressure from both sides. This was true in James Hall's "squeeze box" experiments of 1812, and equally true of Elie de Beaumont's hypothesis of 1852. Again arguing from an overview of the entire Alpine system, Suess maintained that this could not be. If the mountains had not been thrust up from below, neither had they been caught in a vice and squeezed up by the advancing "jaws" of the rigid masses to the north and south of the system. He reasoned that the overthrust of the whole system onto younger beds to the north suggested a one-sided movement, and further, that the Alps described a great wide curve *convex to the north* indicating that they had received a one-sided thrust from the south. If they had been merely crushed in situ there would be no explanation for such a great bowlike flexure of the range.

This conception of the lateral migration of an entire mountain system found no counterpart in any previous theory. It did not conform to an uplift theory, like Studer's, or a theory of symmetrical compression, like Elie de Beaumont's. It was equally incompatible with any Lyellian account of mountains—Lyell's own, Babbage's, Herschel's, or Hall's. Even Dana's theory, which Suess had employed to suggest the relative superiority of transverse over vertical conceptions of orogenic forces, had no place for such lateral movement. It followed previous theories only in its mode of analysis—the trend lines of surface features—and in its general affinity with those nonuniformitarian theories that proposed a decipherable history of mountain dynamics superimposed, on an enormous scale, upon the uniformitarian cycle of erosion and deposition and the regular transgression of the seas. Thus, Suess's work belongs to the same tradition as the theories of Dana and Elie de Beaumont: not necessarily catastrophic, it was decidedly nonuniformitarian.

Suess applied his conception to other ranges as well—the solution of the Alpine problem that he thought he had achieved was a solution that must apply equally to adjacent fold mountains.

> The folds of the Jura describe just such a constant line, unbroken for long stretches; the Jura are quite without [any such] central masses. Not less conspicuous is the contrast between the broken-up central masses and the regularity of the outer folds of the Carpathians . . . which can be explained only by a general and isomorphous movement of the entire mass of the mountains. . . . All these mountains from the Jura to the Apennines and to the Carpathians are distinguished in the same way by their constant predominant fixed lines of strike, and I include all these branches and limbs under the name "Alpine system." [Suess 1875:14]

It was not just the true Alps that had moved forward, *all* the major ranges of Europe had experienced a general northern displacement together and had a connected history as a *system*. Suess imagined the mountains moving forward as waves. These were not the waves of Rogers or Agassiz, thrown up instantaneously by oscillations of the lava-filled cavities of the subcrust; on the contrary, the movement had been imperceptibly slow. That each component of the system described its own curve and that there was not a single linear wave front or a single great curved front for the whole system was the result of the "breaking" of the northern-most waves of the system against the masses of unyielding crystalline rocks in their path. The Massif Central and the Vosges in France and the

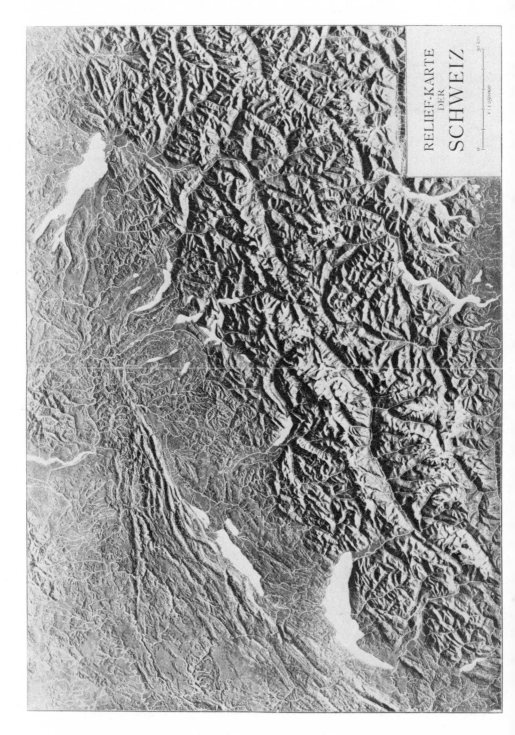

RELIEF-KARTE DER SCHWEIZ

50 km

1:1 250 000

0

[154]

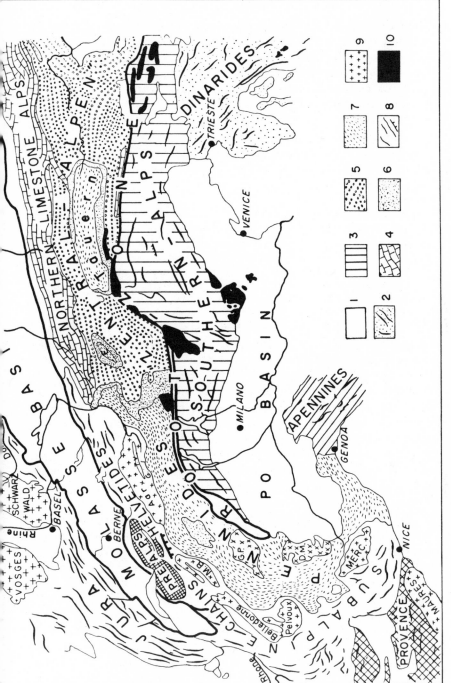

Two maps of Alpine Europe. *Top:* A map of the Alps from Heim 1919, illustrating the major components of the Alpine system, the interpretation of which he shared with Suess. (From Heim 1919–1921: III.) *Bottom:* The major units of the Alpine system corresponding to the map above. Boxed and numbered keys refer to a later interpretation and are not pertinent here. (From Joos Cadisch, "On Some Problems of Alpine Tectonics," *Experientia*, 2 [1946].)

Black Forest and Bohemian Massifs in Germany were the western and northern "shore" or border of the region in which the Alpine system had developed its great regularity. Between each pair of massifs the system had pressed forward in a bow, producing a long series of curves stretching from the Jura to the far end of the Carpathians (Suess 1875:18).

Suess was aware of the enormous implications of this thesis and the extent of his departure from all accepted ideas about mountain ranges, and in completing the sketch of his theory he invoked the name and reputation of Studer to help prepare the way for its acceptance.

> One cannot well deny Studer's opinion that every step in our knowledge of the folds of the northern zone of the Alps compels us to enlarge our view of its cause, and here lies an approach to the idea that has been more widely revealed here . . . that one must accept an extraordinarily great force that has acted from the Alps outward and that does not appear to be directly based on the individual granite massifs. [Suess 1875:16]

While the idea appears to be as improbably schematic as any of its nineteenth-century predecessors, Suess pressed his case well in later chapters of *Die Entstehung der Alpen* with a closer look at each segment of the northern front of the system. Finally, in the closing chapters, he extended the explanatory scheme beyond the borders of Europe to India, Central and East Asia, and North America. If, he concluded, there was no geometrical system such as that envisaged by Elie de Beaumont, there were nevertheless two great series of interconnected mountain trends in the world: all the great fold mountains of Europe and North America pushed in arcs toward the north, northwest, and northeast, and the mountains of Asia advanced in great flexures to the south. Speculating on the cause of such unmistakable regularity, Suess turned to the contraction theory: the relation of all the great mountain arcs to adjacent subsidences suggested that their motions were a response to the unequal radial contraction of a cooling earth (Suess 1875:144).

Suess's work of 1875 marks the beginning in Europe of a new movement toward structural-dynamic studies of the crust as the focus of geological research and toward the "structural synthesis" as the ultimate aim of geology. This movement was in part a result of marked and rapid advances in the identification and mapping of stratigraphical sequences and in the paleontological record since the beginning of the nineteenth century, and the appearance of a work like *Die Entstehung der Alpen* is a measure of the successful application of the fundamental rules of procedure established by Smith, Lyell, Cuvier, and others for tracing the history of life and the depositional history of the continental platforms expressed in the sedimentary record.

Suess's work is a convenient benchmark for a new attempt by geologists to assimilate the results of paleontological and stratigraphical study into a dynamic history of the crust. His work was very much like that of Elie de Beaumont in its major elements, even if Suess came to conclusions somewhat different from those of the great French geologist. Suess's work made trend-line analysis of tectonic features respectable outside of France, where it had become a burdensome orthodoxy. Elie de Beaumont's heavyhanded advocacy of his pentagonal scheme had overshadowed the core of his tectonic ideas and discredited them. Yet the theory of secular contraction, of orogenic episodes, of mountain building through transverse compression, and the analysis of the trendlines of major mountain ranges and their relations, which Elie de Beaumont brought together for the first time, also served as the foundation of the lifework of Suess. Abandoning the laws of geometric intersection, not insisting on the cataclysmic upheaval of mountain chains, departing from symmetrical compression for a more complex interpretation of the transverse forces, Suess reinforced trend-line tectonics (in the context of the contraction theory) with a wealth of new knowledge about the earth's crust gained by the established geological surveys and by independent explorers and geologists in the third quarter of the nineteenth century. The result was a major synthesis of geological work, begun by Suess in 1878, and not completed until 1904. The massive treatise, *Das Antlitz der Erde* (*The Face of the Earth*), eventually filled three large volumes and totaled more than 2,000 pages.

It was not merely the size of the work or the persistence of the author that made the book remarkable. Considering the amount of research digested in its pages it is extraordinary that it is not longer than it is. In fact, the servant who for more than twenty years carried reference works from the geological library of the Hofsmuseum in Vienna to Suess's study exclaimed, on seeing the finished work, "Is that *all* you got out of the books I brought you?" (Hobbs 1914:814). Suess's aim was to compress and integrate the resources of world geology into a single essay on the structure of the earth's crust so that an overview of the whole might be obtained through an understanding of the actual links between structures in time and space across national and continental boundaries. In successive volumes he surveyed the mountain ranges of the earth and the history of the continents and oceans, following the outline plan sketched in *Die Entstehung der Alpen,* treating each of the elements in vastly greater detail. The style of the work was deliberately redundant, stressing the connection of each geographic unit and each geological period to the entire structural history of the crust.

Marcel Bertrand wrote in the preface to the French edition of *The Face of the Earth* that Suess's method "has been able to show the relationship and establish the connections from one limit of our hemisphere to the other, which, for example, had not before been perceived even from one boundary of France to the other. M. Suess has known how to elevate the fundamental features to a sufficient altitude to be seen among the complex details of their surroundings" (Hobbs 1919:134). This superior vantage point was essential to the task Suess had set.

In introducing the argument of *Die Entstehung der Alpen,* it had been sufficient for Suess to transport his audience mentally to the top of an Alpine peak and observe the great arc of the Alps bowing out to the north. But for this larger and more ambitious undertaking, mountaintop views were not sufficient, and Suess chose to have the reader imagine himself to be a visitor from space. The impact of this device is largely lost in an era in which photographs of the earth have been taken by astronauts coming back from the moon and in which beginning geology students are exposed to aerial photographs that reveal at a glance the pictures that Suess took thousands of words to create. But in 1883 the impact was all the greater for lack of such imaginative aids.

In the opening pages, Suess's visitor was induced to push aside the clouds and to contemplate the earth rotating beneath him, to notice immediately the most prominent feature of the surface, "the wedge-like shape of the continents as they narrow away to the South" (Suess 1904–1909:I, i).[8] Had he been able to part the ocean as easily as the clouds and seen "the remarkable depth of the ocean basins as opposed to the trifling height of the continents, and the steep slope of a greater part of the coasts," he would surely not have doubted, mused Suess, that "such forms must consequently

8. All citations are from the English edition (1904–1909) of the German original, which appeared between 1883 and 1904. Whereas the French edition was greatly expanded and illustrated by the translator, Emmanuel de Margerie, between 1897 and 1918, the English edition is an unaltered translation of the original.

have been determined by the structure of the outer part of the planet itself" (Suess 1904–1909:I, i).

Drawing closer to the surface, the observer notes the relations between the continents and the mountain masses they support and sees that the earth has two regions defined by the difference in the relation of the mountains to the coast. The Pacific Ocean is ringed by chains of mountains parallel to the coast, which suggests that "some definite connection unquestionably exists between the outer delimitation of a continent and its internal structure." In the Atlantic region, however, "this rule does not apply" (Suess 1904–1909:I, 5). On both sides of the Atlantic, where mountains approach the sea at all they either turn their backs on it, as in the American Appalachians, or else cut transversely across the continent, as in Scotland, Brittany, and Portugal. Suess's space visitor notes that the Indian Ocean is bordered by both kinds of coast—at the western edge (Africa, Arabia, India, and far to the south, in Australia), there is an "Altantic Coast." The eastern shore—from the mouth of the Ganges to Java—has a coastline of the "Pacific type." "The boundary of the two regions," he concluded, "is to be sought on the mainland" (Suess 1904–1909:I, 6). Indeed, the visitor discovers in the east-west line from the Himalayas across the top of India, through the Persian Gulf, and then across the top of the African continent to Morocco, a rampart of mountains separating the north from the south. Thus, Suess's visitor concludes, "the mightiest mountain chains of the earth are themselves only subordinate members of far greater structural features which dominate the whole globe. We may observe and describe in detail the structure of any single mountain chain we please, but it will still be found impossible to give an explanation of the facts it presents without taking into account the relations which exist between this particular chain and the assemblage of mountain chains in general" (Suess 1904–1909:I, 6).

Suess's observer then set foot on the earth to have a closer look at the mountain chains whose relationship he had noted from far above. Surprisingly, as he walked about, he saw "scant traces of the mighty movements which have affected so many parts of the earth's surface." He saw that other once-mighty chains had been worn down to hills and plains and found that "the fractures, along which displacements of mountain segments have taken place to the extent of thousands of feet, are so completely concealed from the eye that they have only become known at all by means of subterranean workings." These faults, the observer noted, traverse the earth in great systems of fractures (Suess 1904–1909:I, 7).

Now made aware that the nature of the earth can only be understood in the vital connection of the most comprehensive overview and the most detailed study of faults and other hidden structures, the visitor was led by Suess into the universities. There he heard how spectroscopy has extended man's knowledge of celestial bodies and how these bodies are in the process of cooling and how these conclusions have been applied to the study of the formation of the solar system and the origin of life, then to the seas, and to the atmosphere. He was taught the succession of the early geological periods and finally brought to the threshold of stratigraphy and the history of life.

> Surrounded by an overwhelming mass of details concerning the distribution, lithological character, technical utility and organic remains of each subdivision of the stratified series he stops to ask the question: What is a geological formation? What conditions determine its beginning and its end? How is it to be explained that the very earliest of them all, the Silurian formation, recurs in parts of the world so widely removed from one another . . . always attended by characteristic features, and how does it happen that particular horizons of various ages may be compared to or distinguished from other horizons over such large areas, that in fact these stratigraphical subdivisions extend over the whole globe? [Suess 1904–1909:I, 8]

The questions, Suess noted paternally, are obvious and justifiable, but "if we could assemble in one brilliant tribunal the most famous masters of our science, and could lay this question of the student before them, I doubt whether the reply would be unanimous, I do not even know if it would be definite. Certain it is that in the course of the last few decades the answer would not always have been the same" (Suess 1904–1909:I, 8).

Leaving the celestial visitor to his own devices for a moment, Suess sought the reasons for this discord in the history of geology. When he had entered the field, the doctrine of special creations (Cuvier) reigned everywhere as an explanation of the larger divisions of the geological series. Attempts were made to correlate the results of paleontology in Elie de Beaumont's geometrical laws of the distribution of mountains in time and space. Considerable further research indicated that however mountains were raised, the evidence of the slow advance of the seas (and the deposition of marine faunas over wide areas) must force geologists to pay as much attention to transgressions as they had to discordances.

By 1859 it was asserted that the dislocation of the crust did not determine the succession of life, which was governed instead by slower oscillations far greater than the local elevation of mountains. Even in France the theory of repeated and sudden destructions of life had fallen into disrepute, attacked in a variety of ways by new findings. By 1859 the majority of investigators attributed the differences presented by successive sedimentary series and their faunas to the slow and extensive oscillation of continents and to repeated climatic changes. The appearance of Darwin's *On the Origin of Species* bespoke further evidence of some great slow process; the lack of intermediary steps that could trace the linked evolution of all forms was attributed by Darwin to the imperfection of the geological record. Research since Darwin, though it had filled in many of the gaps, still revealed the existence of entire populations of animals and plants that came into being together and disappeared together. It became more and more clear that "the determining factors . . . have been changes in the external conditions of life" (Suess 1904–1909:I, 11). This view was reinforced when geologists traveling to Australia, India, China, and the Americas were able to make unhesitating use of the stratigraphical terminology chosen to classify deposits in limited parts of Europe, and the study of sedimentary cycles was increasingly applied to the solution of the history of continental oscillations.

However important such sedimentary cycles might be, they were for Suess no solution to the problem of continental movement. While there was no longer any question that the analysis of single chains of mountains (considered in a worldwide context) could be the answer to the transgression of the seas upon the continents, it appeared to Suess equally unlikely that whole continents could be smoothly elevated and lowered without relative dislocation of parts. Moreover, "the manner in which the contraction of the earth's crust manifests itself on the surface of the planet, in the formation of folds and faults, does not accord with the hypothesis of moving continental masses, which, over wide areas, repeatedly ascend and descend in a slow and uniform manner" (Suess 1904–1909:I, 14). It is this opposition, Suess concluded, between continental oscillation and secular contraction, that would lie at the root of the discord that the celestial observer would meet in the universities.

Finally, let us suppose that our imaginary observer, wanderer, and listener should now leave the lecture room, and seek instruction as to the true nature of a geological formation in our abundant literature. If he thought it worthwhile to open the book which I now offer to publicity, he would not find in it an answer to his question. This answer is the great task of the next generation of investigators. Here we will only attempt by critical synthesis of new observations to dissipate many ancient errors and to prepare the way for an unprejudiced survey. [Suess 1904–1909:I, 15]

Suess's "observer" was a convenient means by which to introduce several generalizations that dominate the work from beginning to end. The differentiation of the Atlantic and Pacific provinces of the earth and their graphic separation by a globe-circling mountain system was fundamental to Suess's interpretation of the history of the continents and oceans and their connection with the history of the earth as a cooling planetary body. Further, the mountain systems that were for so long the "key" to the understanding of the structure of the earth now provided that key only when understood as subsidiary elements of even greater units of analysis—the continents. The history of these greater units was not written only in the strike of *existing* mountain chains—closer examination revealed that other great mountain systems have been born and have perished on the continents and they, too, tell the story. For Suess, relief was not always the key to structure—one must tie in the remains of nearly vanished structures with those that now exist if one is to grasp the connected history of the globe.

In *Die Entstehung der Alpen* Suess had employed a synthesis of new observations to combat the theory of the plutonic uplift of mountain ranges. In *Das Antlitz der Erde* a much more extensive synthesis was employed to oppose the theory of continental oscillations, which, he argued, failed to explain the facts of dislocation and was not in accord with the theory of secular contraction. The argument he presented in the introduction is very much like that used by Dana against Hall, only here the word *mountains* must be replaced by *continents*: the theory of continental oscillations is a theory of the origin of continents with the origin of continents left out; the structural peculiarities so evident to the space visitor from continent to continent—the linked mountain system, the networks of fractures, the differentiation of Atlantic and Pacific coastal types, the circumglobal line of demarcation which divides the north from the south, the shape of the continental masses, all are unexplained by a theory of uniformitarian oscillations and sedimentary cycles.

In the brief historical summary of geology in his own lifetime Suess noted the inability of catastrophic theories of marine transgression to give a plausible explanation of the marine faunas on the continental platforms, either of their spatial distribution or of the great thickness of the strata containing them and the great lengths of time this implied. Suess was also aware that the theory of evolution proposed by Darwin accorded more with uniformitarian conceptions of the alternation of land and sea on the continents than with opposing theories of special creation and epochal disturbance. He was equally aware of the importance of Lyell's geology in providing an adequate time scale for Darwin's evolutionary theory. Since, as we have seen, studies of mountain ranges were generally carried out by geologists who espoused either catastrophic or nonuniformitarian conceptions of geological change, the success of Darwin, and of Lyell, had recently caused a general turning away from these studies as somehow tainted with an outmoded and inadequate view of the sources of geological change and its time scale in particular. This is what Suess particularly wished to undo.

In the opening chapter of *Principles of Geology* Lyell had argued that the polemical excesses of the Huttonians and Wernerians had caused a long-term aversion to all theory on the part of geologists. In Suess's view, the excesses of Lyellian uniformitarianism had recreated that state of affairs in the 1880s.

> The enthusiasm with which the little polyp building up the coral reef, and the raindrop hollowing out the stone, have been contemplated, has, I fear, introduced into the consideration of important questions concerning the history of the earth a certain element of geological quietism—derived from the peaceable commonplaces of everyday life—an element which by no means contributes to a just conception of those phenomena which have been, and still are, of the first consequence in fashioning the present face of the earth.[9]

Whatever great service Lyell had performed in showing how "Nature may attain striking results by trifling means," nonetheless, the standard for great and small, for long and brief, in the evaluation of geological change had been obtained by an illicit and erroneous anthropomorphic reduction. "We often measure mountains in feet, and we distinguish long and short periods of time

9. Suess 1904–1909:I, 17. Suess's use of the term *quietism* pokes fun at the pious contemplation of the majestic cadences of geological change in the uniformitarian system.

according to the average length of human life, that is, according to the frailty of our bodies; and in like manner we unconsciously borrow the standard for the terms 'violent' and 'less violent' from the sphere of our own experience. . . . we are prone to forget that the planet may be measured *by* man, but not *according to* man" (Suess 1904–1909:I, 17).

The coral polyp and the raindrop chosen by Suess to exemplify uniformitarian excesses were not random examples. The theory of coral reefs put forward by Darwin was a principal piece of evidence for the slow and general subsidence of the floor of the Pacific suggested in the uniformitarian theory of continental oscillations; the raindrop evoked the importance of subaerial denudation in the Lyellian explanation of the irregularities of surface relief and the mechanism of geological change. Certainly these phenomena existed, but like the use of our size and life span to measure the world, they were not the measure of the whole.

> The convulsions which have affected certain parts of the earth's crust, with a frequency far greater than till quite recently supposed, show clearly enough how one-sided this point of view is. The earthquakes of the present day are certainly but faint reminiscences of those telluric movements to which the structure of almost every mountain range bears witness. Numerous examples of great mountain chains suggest by their structures the possibility, and in certain cases even the probability of the occasional intervention in the course of great geologic processes of episodal disturbances, of such indescribable and overpowering violence, that the imagination refuses to follow the understanding and to complete the picture of which the outlines are furnished by observations of fact. [Suess 1904–1909:I, 18]

This assertion is Suess's position in a nutshell. It did not denigrate the efficacy or even the predominance of the secular work of erosion and sedimentation—these are the "great geologic processes" to which he alluded. The one-sidedness of uniformitarianism, as he characterized it, was a philosophical predisposition to explain away, or ignore, observational evidence of violent, catastrophic disturbance of the crust. Geology was for Suess an act of imaginative reconstruction of the arrangement of things before they were disturbed, and the uniformitarians suffered from a failure of the imagination required for such a task. That the great disturbances are principally evident in great mountain ranges, and that the lack of a theory of mountain ranges is a substantial gap in Lyell's work, are two sides of the same coin.

The principal problem faced by nineteenth-century geology as Suess saw it was a limitation of perspective. If we agree to look at mountains like the Alps *as they stand,* the overwhelming first impression is the work performed by erosion and weathering. The peaks are etched by frost, and the arrangement of peaks and valleys in systems is conditioned by the way water has flowed away from the peaks. But a deeper look at, for example, dislocations along fault lines shows a different set of phenomena at work—episodically, but no less dramatically in its effects.

Catastrophism and uniformitarianism for Suess were the extreme end results of extrapolation from two different sets of phenomena superimposed on one another. In Buch's geology, erosion did not even exist, and all relief was the expression of "great telluric movements." Lyell conversely seemed to ignore the evidence of greater movements in favor of the continuous action of small causes. Neither position adequately describes what we see when we look most carefully. Lyell's version is true most of the time, but episodic disturbances have a role much greater than might be inferred from their rarity and what we have seen of them. Again, it is the use of our life span to measure rarity or regularity that is at fault, and the same holds true for the use of our size to measure the extent of geological disturbance. "Such [great] catastrophes have not occurred since the existence of man, at least not since the existence of written records. The most stupendous natural event for which we have human testimony is known as the Deluge" (Suess 1904–1909:I, 18).

It was the overthrow of the deluge notion that was the first success of uniformitarian geology. "One thing the *Principles of Geology* unquestionably accomplished," wrote Gillispie. "The book administered the coup de grace to the deluge" (Gillispie 1959:ch. 4). This makes it all the more striking that Suess, writing in 1883, should devote the first chapter of a book like *The Face of the Earth* to a reconsideration of the deluge. It is striking not because Suess raises a question bound to excite strong opinions and awaken old antagonisms; this might be expected as a part of his attempt to "dissipate ancient errors and to prepare the way for an unprejudiced survey." Rather, it is because Suess emphatically defended the reality of the deluge against the quietistic excesses of uniformitarian geology.

Yet Suess was no biblical geologist and not a catastrophist of the old school. As his study unfolds, it becomes increasingly clear that he deliberately chose the most emotional and disputed ground of an old antagonism as a showcase for his own approach to geology, and to bring an end to the debate forever. Suess intended a tour de force in which all the resources at his command would be brought to bear on a sensitive question that would give a new content to the term *uniformitarian* and suppress the parochial dogmatism that inspired the most emotional resistance to it. Chastizing geologists for a too-great fascination with minute causes, he gave ample testimony that catastrophes as great as the biblical deluge dotted recorded history. On the other hand, he applied both geological science and historical and archaeological research to show that the biblical account was derivative from older sources and that the deluge was a Mesopotamian event and not a universal one. Both camps were thus neutralized. He admitted a catastrophe, but the "wrong" one, from a dogmatic standpoint. Neither side in the old debate could be much comforted by his conclusions.

The problem of the deluge was a warm-up exercise for Suess, and an ingenious one. Remembering what he said about the study of a single mountain range revealing little without the study of those related to it, we see a corollary here. Devotion to a single text (the Bible) or to a single philosophy (uniformitarianism) can only lead us astray—the true nature of the deluge can only be illuminated by the study of all aspects of all the evidence at our disposal. When such an investigation is carried out, old antagonisms born of one-sidedness fall of their own weight. If the tendentious history of geology in Lyell's *Principles of Geology* reveals the strategy of that work (Rudwick 1970), Suess's examination of the deluge is certainly no less the key to the strategy of *The Face of the Earth*.

Suess's aim was to enlarge the meaning of uniform change to include phenomena not presently observable. While he obviously had in mind the tremendous dislocations evident in mountain systems, he began more modestly with a consideration of a catastrophe greater than any seen in modern history, yet one for which we have *written records:* eyewitness testimony of those who experienced the phenomena associated with the flood. This is not a strategy that could compel assent to the idea of the great northward movement of the Alps suggested in his earlier work, but it made a promising start: if successful, it could enlarge the notion of "causes presently at work" to include events through thousands, not mere hundreds, of years, while remaining within the frame of documentary evidence and direct observations.

CHAPTER SEVEN

The Face of the Earth:
Eduard Suess and the Second Global Tectonics

Suess had a variety of aims in returning to a consideration of the deluge at the beginning of his major work. If Lyell had over-thrown the biblical geologists in England, the arguments by which he had done so appeared regressive in the late nineteenth century. Lyell had argued that no particular series of formations could be associated with or correlated with the specific events mentioned in the text of the Bible. Gaining general assent on this point, he used it as an attack against all geological catastrophes and as an argu-ment in favor of his own uniformitarian philosophy of geology. In hanging a philosophical postulate on the argument from ignorance of which geological phenomena correlate with the deluge, Lyell suggested the logical avenue of attack—if these phenomena could be located, and the deluge given a concrete, nonsupernatural inter-pretation, then the geological quietism fostered by Lyell's sublime contemplation of the majestically slow cadence of change might be undone, and the idea of intermittent and more violent change might once again achieve theoretical respectability. But in turning to the written evidence for the existence of something like the deluge, Suess bypassed the geological improbabilities of the pluvial deluge of Genesis and turned instead to the recently discovered cuneiform text of the heroic epic of Izdubar, recovered from the library at Nineveh (Suess 1904–1909:I, 20–21). In so doing, Suess achieved a second strategic aim—the debate must finally be re-solved, like all scientific questions, on the basis of the evidence. Suess strove everywhere to give his work the sense of something new and up-to-date. *The Face of the Earth* was not intended as a beginning text for students but as a mature summary of geology written by a professional geologist for his peers; it could receive the attention he desired for it only if it reflected the most recent and best geological research. It therefore was necessary to approach the question of when and how something like a great flood might have happened by appeal to a text both "newer" and older than the Bible. As he avoided the absolutist oppositions of the old debate on the geological side, he would also refuse to be wedded to special pleading for any particular catastrophe based on an attempt to shore up religion with geology. The account of the flood in Genesis, with the earth inundated by forty days of rain, and the ark coming to rest on a high mountain peak, was both historically derivative and geologically improbable; the cuneiform text was preferable not only because of its greater age but because the phenomena to be deduced from it were more reasonable from a geological standpoint.

Suess began his quest for the deluge in a textual examination of his cuneiform epic, with enough critical discussion of alternate readings in various translations to assure the reader that he had fully mastered the subject and had not fallen into the dangerous error of copying another's work unexamined. That completed, he began to interrogate the text as an archaeologist and a geologist, dusting off every hint of the fragmentary evidence. That the ark of Genesis, and the boat of the hero of the cuneiform epic are both smeared "within and without" with pitch is sufficient for a chain of inferences that already locates the site of the true deluge. The Miocene strata of the Tigris-Euphrates valley are rich in asphalt. A report of a recent railroad construction party there contains the observation that boats are constructed by the inhabitants using pitch "within and without." Both the biblical story of the tower of Babel and Herodotus in his description of Babylon noted the use of pitch as a mortar for brickwork. Reference is made in cuneiform texts to the use of bitumen in fire bombs (Suess 1904–1909:I, 26–27). Already, in the first pages of the work, classical and biblical scholarship and the history of warfare are brought together to pinpoint the deluge as a Mesopotamian event.

Suess treated all aspects of the epic in detail and in the same pattern. A phenomenon is noted in a fragment of text. That phenomenon is then associated with some geological phenomenon noted in the present, and the geological literature of the world is culled to establish its actuality, its range, and its frequency. One example of this procedure will suffice, a treatment of a scant four lines of the cuneiform epic (Suess 1904–1909:I, 30).

47 The Annunaki cause the floods to rise
48 The earth they make to tremble through their power
49 Ramman's great billow ascends to the sky
50 All light is consumed in (darkness)

Proceeding line by line Suess was able to reconstruct the entire deluge from this passage. The Annunaki of line 47 are the spirits of the deep, therefore the floods referred to must come from the ground, as opposed to the sea or from rain. Specifying the nature of the phenomenon, Suess moved immediately to the geological interpretation.

. . . the rising of great quantities of water from the deep is a phenomenon which is a characteristic accompaniment of earth-quakes in the alluvial districts of great rivers. The subterranean water is contained in the recent deposits of the great plains on both sides of the stream, and . . . what lies beneath this limit [the mean level of the river] is saturated and mobile; the ground above it is dry and friable. When seismic oscillations occur in a district of this kind the brittle upper layer of the ground splits open in long clefts, and from these fissures the underground water, either clear or a muddy mass, is violently ejected, sometimes in great volumes, sometimes in isolated jets several yards high. [Suess 1904–1909:I, 31]

The plausibility of this reconstruction was reinforced by Suess's wide command of the literature of geology. Such upwelling of water was observed in the Danube delta in 1838 and 1879, and in the Mississippi Valley in 1812. Suess mischievously included, not quite appositely, an observation of Lyell's concerning the draining of a lake through bottom fissures during an earthquake. Numerous further examples were adduced from Siberia and from the lower reaches of the Indus, Ganges, and Bramaputra rivers in India. Finally, he concluded that "these floods . . . offer in my opinion a conclusive proof to the geologist that we are here dealing with a seismic convulsion in a broad river valley. Such phenomena have never been observed on any great scale except in extensive low-lying districts . . . nor would they be explicable under any other conditions" (Suess 1904–1909:I, 33).

We see the same plan of attack followed for line 49, the frag-ment "Ramman's great billow ascends to the sky." Ramman is not the god of the sea but of weather, and the physical phenomenon represented is the arrival onshore of a violent cyclonic storm, a hurricane.

> Sudden and terrible are the floods caused by cyclones. They only occur in proximity of the sea, either on islands or over the plains bordering the lower courses of great rivers. . . . if [the storm] is checked by the narrowing of the sea between convergent coasts, it rises to a steadily increasing height until it finally precipitates itself over the plain, carrying destruction before it. [Suess 1904–1909:I, 33]

The confusion by primitive peoples of this phenomenon with a pluvial deluge (as in the Bible) is understandable, said Suess, since the actual flooding is both preceded and accompanied by torrential rains and violent thunderstorms. Then the recourse to current observations was repeated: these phenomena have been witnessed in Calcutta in 1737, in the West Indies in 1780. Further, in both cases an earthquake accompanied the flooding. The darkness was a part of the cyclonic storm, and the whole assemblage was part of a single known phenomenon. "The most destructive phenomenon of the present day, the cyclone, accompanied by an earthquake is also that which corresponds most closely to [the] description of the greatest natural event of antiquity (Suess 1904–1909:I, 34).

This clearly was the major point of the entire exercise, but Suess pressed on relentlessly. The duration recorded corresponds to that of modern hurricanes, and the phenomena associated with the withdrawal of the water and the stranding of the heroes' boat *inland* are further proof that the deluge was the result of the phenomena described. "The fact that the vessel was driven inland from the sea against the course of the rivers seems to me decisive of the character of the entire catastrophe. If, as is generally supposed, the flood had been caused by rain, it would certainly have carried the ship from the lower Euphrates to the sea (Suess 1904–1909:I, 38).

Having accomplished this reconstruction, Suess turned from his Mesopotamian epic to "discuss events of our own times similar to that of the Deluge. The accounts of these in the last twenty or thirty years show that they are much more frequent than we, owing to our European position, are apt to assume" (Suess 1904–1909:I, 40). These more recent events were the inundations of the deltas of the great rivers of India, including at least two occurrences of simultaneous earthquakes and cyclones in these areas. These and numerous additional descriptions of similar events had the obvious effect of establishing the nature of the physical agencies involved and their regularity in the areas concerned.

After all of this one might expect Suess to have rested his case, but he returned refreshed from this excursion to argue another side of it. This time his target was not the uniformitarians who would deny such a catastrophe but the catastrophists, who would make of this event a universal phenomenon and geological alkahest. Considering again the biblical and Mesopotamian accounts, he searched out Egyptian, Hellenic-Syriac, Hindu, and Chinese accounts of great deluges. He dismissed the possibility that these were faint recollections of a single universal deluge on various grounds: faulty or nonsynchronous dating of the texts, the derivative nature of some of the accounts, and the relative likelihood that each of them could be economically explained as often repeated events appropriate to the geological and climatological conditions of each region.

Finally exhausting his ammunition in this direction, Suess restated his conclusion. The deluge was a local Mesopotamian event in which a seismic and cyclonic disturbance of great violence caused the flooding of the plain of the Euphrates from the sea, and in which, as recorded in the legend, a man with enough foresight to get himself and his family into one of the pitch-caulked boats characteristic of the region was washed far inland where, as the flood abated, he was "stranded on one of those Miocene foothills which bound the plain of the Tigris on the North and Northeast below the confluence of the little Zab" (Suess 1904–1909:I, 72). This matter-of-fact anachronism, in which Hasis-Adra, the Mesopotamian philosopher, lands on *Miocene* foothills, is the most compact statement of the relationship of the epic to modern geology: for this scientific time designation wrenches the story out of the half-light of mythology and renders it part of history.

Not the least of Suess's aims was to establish his own mastery, before his audience, of the widest imaginable spectrum of materials. From Assyriology to Egyptology to the anthropology of the Pacific, from political history to travelers' and missionaries' records, geological surveys, the records of colonial officials, sinology, meteorology, classical scholarship, geology and geography, and the records of voyages and explorations ranging over centuries, he moved effortlessly from subject to subject. Yet from this seemingly disparate material he assembled the case for the idea that from the Mediterranean to the Yellow River the world has been visited repeatedly by catastrophes greater than anything known in the recorded history of Europe, including the infamous Lisbon earthquake of 1755. All of this drove home the fundamental message: any theory based exclusively or even primarily on the geography, climate, and history of a single region had to underestimate vastly the actual scale of current catastrophic alterations of the crust of the earth and to minimize the range and power of such catastrophes in general. Having recast and confirmed the story of the deluge, Suess was no longer shy about inferring greater and (in human time) rarer catastrophes at some time in the geological past from the existence of contemporary catastrophes and the records of a very great catastrophe barely within the horizon of human history. Simply because Europe does not now experience, and has not for ages, the great telluric movements that made the Alps into the great deformed structures they are today, there is no presumptive reason to dismiss the possibility that such a thing has ever happened.

On the other hand, Suess went to great lengths to dispel the notion that there was ever a single universal event in geologic history preserved in the Bible and the records of other ancient civilizations or in the oral traditions of preliterate peoples. The Mesopotamian deluge was one of many discrete, local, violent events. Hundreds of these dot human history and they are part of the regular march of geological change in which faulting, flooding, sudden alteration of river courses, construction of natural dams, and catastrophic sedimentation are regular, *occasional* processes superimposed on the cycle of erosion and sedimentation. But they are no less a part of the subject matter and formative structure of geology than the secular processes to which the uniformitarians too exclusively devoted themselves. When, with disregard for "rarity" measured in human time, these occasional processes have occurred jointly, the effects have been catastrophic—but they have never been universal, nor occult in nature.

It was a sound argument, well constructed and well aimed. But uniformitarian geology was not shored up merely by clever debating tricks, and neither could an alternate theory of the earth find foundation in a causerie, however pleasant, concerning the geological riches of ancient texts. The real work at hand was to provide a theory of the dynamic history of the crust as powerful as that generated by Lyell in the many editions of *Principles of Geology*. In the chapters following his consideration of the deluge, Suess went methodically to work to provide such a theory. The remainder of the first part of *The Face of the Earth* consisted of four chapters under the general heading "Movements in the Outer Crust of the Earth." Earthquakes, faulting, dislocation and disruption of sediments, volcanoes, and other igneous and metamorphic phenomena were considered in turn, and the diversity of their interconnections established. The geological results of their concerted action in different geological provinces were examined with an eye to the complex causality of even the most regular terrains.

Such consideration was a necessary prologue for the detailed analyses projected for succeeding volumes of the work, in which the whole face of the earth was to be examined. Even in this technical discussion of geological dynamics, however, the strategic and polemical motives, which marked the reconstruction of the deluge, remain everywhere evident.

All the material introduced in these technical chapters attacks the uniformitarian doctrine of continental oscillation and the possibility that the crust of the earth experiences significant upward movement of any kind. On the other hand, it supports the notion that crustal movements are either the direct effects of subsidence as the earth cools and contracts or the horizontal resolution of stresses encountered by the crust as a result of this same contraction. No one has argued that Suess was not wedded to the theory of secular contraction, but it is generally not emphasized that *The Face of the Earth* was put forward as an integrated defense of that hypothesis as opposed to the theory of continental oscillations: he presented evidence of the great horizontal displacements of the crust in the overfolded mountain belts that he believed formed the outer borders of regions of subsidence. Although later geologists argued that Suess's descriptive mastery was not fatally entangled with the "separable metaphysics" of the contraction theory, one must nevertheless appreciate the extent to which the first great synthetic work of modern geology served to advance a particular conception of earth dynamics. This is especially true if we are to understand the crisis experienced by the world geological community when that conception began to meet formidable opposition at the turn of the twentieth century from increasingly sophisticated work in geophysics. The emergence of an international consensus beginning in the late nineteenth century around the work of Suess was also the emergence of a consensus concerning the general superiority of the contraction theory over the oscillatory model of the uniformitarians. When the theory began to fail, the consequences were profound. However, without carrying further this anticipation of later argument, a return to consideration of the introductory chapters of

The Face of the Earth can demonstrate the central conceptions of this emergent synthesis, Suess's dedication to the establishment of the study of horizontal movements of the crust as most important for geology.

Fundamentally, Suess's work argued that dislocation of the crust is ultimately an expression of subsidence. "The dislocations visible in the rocky crust of the earth are the result of movements which are produced by a decrease in the volume of our planet" (Suess 1904–1909:I, 107). This applies alike to horizontal and vertical movements of crustal segments.

> The tensions resulting from this process show a tendency to resolve themselves into *tangential* and *radial* components, and thus into horizontal (i.e., thrusting and folding), and vertical (i.e., sinking) movements. Dislocations may therefore be divided into two main groups, of which one is produced by the more or less horizontal, and the other by the more or less vertical displacement of larger and smaller portions of the earth's crust. [Suess 1904–1909: I, 107]

Thus stated, the position might encompass almost any version of the contraction theory proposed from the time of Leibniz and Descartes to Suess's own day. What distinguishes Suess's work from earlier attempts at a theory of earth contraction is that he developed regular patterns of subsidence with characteristic features through comparative regional studies. Earthquakes, volcanoes, and horizontal displacements received similar treatment, resulting in a series of typologies of earth movement exhibiting characteristic features. These typologies must be reconstructed from the flow of description. *The Face of the Earth* differs from other theoretical treatises in that it is not a series of didactic statements or principles interspersed with occasional examples drawn from field geology, but the reverse: the repetition of descriptions is only occasionally interrupted by a phrase, sometimes italicized, sometimes not, that contains the conclusion to be drawn.[1] The style of argument throughout echoes that of the treatment of the deluge in the opening chapter—recursive and continuous. The effect is twofold: the book cannot be dipped into, it must be read straight through, since crucial passages are buried in the masses of descriptions; each generalization thus appears to be a pure induction from phenomena rather than the application of a principle, derived from limited instances, to the manifold of occurrences.

1. This distinguishing characteristic of Suess's work was accentuated by Sarton, who compared Suess's "true synthesis" to the "didactic synthesis" of Emile Haug's *Traité de géologie* (1907–1911) (George Sarton 1919:382).

While a didactic reconstruction does some violence to Suess's strategy of presentation, the only alternative is to follow him from mountain to mountain, which is impossible here. Suess developed a typology that was originally based on an "Alpine-Mediterranean" model, and he extended it by analogy to other portions of the crust. In this model the contraction of the earth continually separates the crust from the interior as the latter cools and shrinks. The crust is continually strained, leading to both tangential thrusts and vertical subsidences as a resolution of stress (Suess 1904–1909:I, 170). Where the resolution is predominantly vertical, the pattern of faulting and subsidence is circular or polygonal: a general peripheral fault is linked by radial faults, less easy to trace, to a center of subsidence. Alternatively, vertical subsidences may, where horizontal pressure is completely absent, be resolved along lines of fracture following no regular pattern: these are called zones of diverse displacement (Suess 1904–1909:I, 32).

In regions where the resolution of stress is predominantly horizontal, the earth is compressed into long folds. A study of mountain chains reveals a tendency for the anticlinal ridges of this fold structure to be overfolded in the direction of primary movement and for the overfolds to be split at the median axis and overthrust —a structure visible in the Appalachians, the Alps, and the Jura (Suess, 1904–1909:I, 110–111).

Where subsidence and tangential thrust are combined, as when a fault line parallels the strike of the chain, the folding and forward advance of the transversely compressed mass is greatly exaggerated when the foreland subsides: "the horizontal movement becomes far greater, as if it had been assisted by the subsidence" (Suess 1904–1909:I, 141). In addition to the tendency for the folds to overthrust a foreland, there is a tendency for the area within the curve of an advancing fold range, the "backland," to collapse under the strain of the lateral pressure and form a "cauldron subsidence"; this can be observed in the Alps, the Apennines, the Taurus, and the Caucasus (Suess 1904–1909:I, 133–138). This backland subsidence is characteristically accompanied by the extrusion of lavas. The separation of the crust from the interior creates cavities or "maculae" that fill with lava; when the strength of the crust is overcome and a subsidence occurs within the ramparts of a folded range, the lava is extruded along fissures and faults created during the subsidence—volcanism is an epiphenomenon of subsidence (Suess, 1904–1909:I, 170).

These schematic elements can be recombined to describe the history of the Alpine region of Europe. The region that now forms the northern foreland of the Alpine system is a "rigified" remnant of an earlier epoch of mountain building and destruction. The crystalline massifs of the Vosges and the Black Forest and Bohemia are *horsts*, masses of the crust left standing during an epoch of downfaulting or "polygonal subsidence" in which large plates of the crust subsided into the interior of the earth. The major faults are seen in the Rhine-graben and the trough of the Danube, which today cut through the Alpine folds. While stresses caused by the contraction of the earth accumulated, the present Alpine system began to develop great folds and move toward the north. Where there was profound subsidence (to the northeast) the Carpathians advanced unobstructed and overthrust the Russian platform for great distances. Resolution of the stresses driving the Alpine system to the north was achieved in the subsidence of backlands—the Adriatic and the western Mediterranean. In the latter, the volcanism characteristic of such a backland collapse onto lava-filled maculae between the crust and interior of the earth is expressed along the coast of Italy from Naples to the Straits of Messina.

This picture of the most recent evolution of Europe was supported in *The Face of the Earth* by some of Suess's most striking reconstructions. To substantiate the northward movement of the Alps he reviewed accounts of major earthquakes back to the sixteenth century, pointing out that they were all propagated at right angles to the strike of the Alpine chains, that they followed defined lines, and that they were all transmitted further into the northern plain than to the south (Suess 1904–1909:I, 81).

The great overthrusts of the Carpathians were even more dramatically confirmed. The Russian platform is a series of Paleozoic strata from Finland to Southwestern Russia, nearly continuous throughout its extent. The entire series ends abruptly at the eastern border of the Carpathians. Suess's postulation of a great overthrust of the Carpathian folds onto the Russian platform was verified when subsequent prospecting, initiated on his advice, confirmed his prediction that the coal measures of Silesia would be found beneath the Carpathian folds (Suess 1904–1909:I, 188).

The concept of backland "cauldron-subsidences" with attendant volcanism was supported by a sifting of reports of earthquakes and volcanic eruptions, which created a picture of a shallow "dish" of subsidence, such as that centered on the Lipari Islands to the east of the Straits of Messina, exhibiting radial fractures emanating from a volcanic center (Suess 1904–1909:I, 93).

These particularities of the Alpine system do not, however,

stand as an isolated case in Suess's reconstruction of the history of the crust from the elements of subsidence and overthrust. Discussion of the cauldron subsidence of the Lipari Islands leads into a discussion of the volcanoes of Central America, migrating along lines of fracture toward the Pacific coast. This leads in turn to a discussion of observations on the *supposed* (from Suess's point of view) spasmodic elevation of the Andean coast of South America, chronicled by Darwin. Suess's championing of his own views is linked at every turn to the repudiation of the Lyellian account of the creation of mountains, the source of earthquakes and their effects, and the meaning of volcanism as the random and cumulative expression of uniformitarian processes. Suess saw Darwin's observations as the only reasonable alternative to the contraction theory developed in the nineteenth century and took pains to show that the "elevations of the strand" in western South America (the evidence of former beaches one above the other) was a misinterpretation by Darwin of a variety of processes—changes in currents, filling of bays with sediments, and so on, better explained by the ensemble of subsidences and local accommodations in Suess's own arsenal of tectonic forms and causes.

Nowhere in the catalogue of basic "tectonic elements" was room provided for uplift of any but the most limited and transitory kind. Moreover, each element represented the fulfillment of a tendency, a long accumulated stress, an ongoing process with a predictable outcome, all of which cut at the philosophic foundations of the uniformitarian position. The exclusive concentration on the sequence and interrelation of dislocations causes the work of the polyp and the raindrop to fall ever further into the background: erosion and sedimentation seem more and more a stage on which the earth plays out a drama of infinitely greater import. "By the tangential movement, those long folded ranges are produced which traverse the continents from end to end. . . . in spite of the extremely great length of the folded ranges they have been obliged in their local development, and especially in their outer border, to accommodate themselves to pre-existing conditions (Suess 1904–1909:I, 604). Where the folding has taken place in restricted areas, subsidence leaves its mark everywhere—as tableland troughs, peripheral sinking of plateaus, cauldron subsidences on the inner borders of folded mountains, subsidence of folded mountains along transverse fractures. The character of the vertical movement varies greatly, but the cumulative effect is enormous: "It is to subsidence and collapse that the Mediterranean seas and the larger oceans owe

their origin and enlargement." The continents that we investigate
are only the accidents of the collection of oceanic waters in areas of
great subsidence, and these remnants will someday sink in their
turn: "The breaking up of the terrestrial globe, this is what we
witness" (Suess 1904–1909:I, 604).

It is small wonder that the French geologist and historian Louis
De Launay identified Suess with a return to a more cataclysmic
geology. In his history of geology, published in 1913, De Launay
noted that French geology in the 1870s had been, briefly, uniformitarian and had overthrown the doctrines of Elie de Beaumont.
However, he wrote, "since 1880 the predominant influence of M.
Suess has again made the word 'cataclysm' acceptable in scientific
circles . . ." (De Launay 1913:88). De Launay recognized the links
between the work of Suess and Elie de Beaumont, particularly that
vast syntheses of geology "have their solid and necessary basis in
tectonics" (De Launay 1913:231). Recognizing that Suess had repudiated basic aspects of Elie de Beaumont's work, he saw them
nevertheless as part of the same tradition—for one thing, no one
between Elie de Beaumont and Suess had dared to attempt a synthesis of geology; moreover, they were both part of a longer tradition that included both Buch and De la Beche. De Launay did not
approve of this tradition, but he was aware of its origin: "Among
all the mistaken ideas of Werner, that of searching in the accidental
forms of the terrestrial crust for a precise geometric alignment has
had a particularly great success among his disciples" (De Launay
1913:79–80).

Suess recognized himself as a representative and defender of a
long tradition, and the genealogy suggested by De Launay is amply
confirmed by Suess in a variety of writings, for Suess himself saw
four great steps in the history of geology. The first was the work of
Werner, who pioneered study of the manner of deposition of formations and the systematic study of districts. Next came that of
Cuvier with his discovery of classes of life forms coming into and
going out of existence; allied to this was the third step, of William
Smith: the identification and correlation of strata by their fossil
contents. Last was Buch's discovery that mountain structure was
the key to the whole of the science of geology (Suess 1897b:273).
Elie de Beaumont's theory of mountain ranges was "for one reason
alone no longer tenable"—mountain chains are arcs, not straight
lines (Suess 1897a:275). Suess compared his own reconstructions of
vanished mountains from isolated fragments to the anatomical reconstructions of Cuvier, which had inspired him as a young man

(Suess 1904–1909:I, 267). De la Beche's influence had been equally crucial:

> Here I follow principally the description given as early as 1846 by De la Beche; this I cannot mention without an expression of deep gratitude to the author, now long since dead, since it exercised many years ago a decisive influence on my own views as to the structure of great mountain ranges. Although published forty years ago, this description is inspired by those conceptions of the formation of mountain ranges by lateral pressure, and of the true influence of granite masses, which are now winning their way step by step to general acceptance [Suess 1904–1909: II,84].

The general acceptance of the theory of lateral pressure in the creation of mountain ranges was one of the goals of the work upon which Suess had embarked; another was the establishment of the connections between the systems of laterally folded mountains. In the remainder of the first volume, Suess discussed the great mountain systems in terms of his "foreland—folded rampart—backland" scheme of subsidences and tangential thrusts. He chronicled the unidirectional motion of the mountain folds in the systems—in North and in South America to the west, in Europe to the north, and in Asia to the south. He spent a great deal of time specifying the relations between the mountains of Europe and of Asia, particularly the reasons for the reversal of direction of the folding in the Balkans, which connected the north-moving Alpine system to the south-moving Asian ranges. "In no other mountain chain is such a peculiar two-fold expression of the tangential movement known; and here we are precisely in that region in which the limit between the Asiatic movement to the south and the European movement in the opposite direction must be sought" (Suess 1904–1909:I, 490).

The linkage of the Eurasian mountain system was important for Suess's theory because it marked the "true" border of the actual "organic" units of the continents. From North Africa through the Mediterranean to the valley of the Tigris and the Persian Gulf, up the course of the Indus, around the Himalayan arc, south through the Bay of Bengal, looping eventually to Sumatra and Java, "the whole Southern border of Eurasia advances in great folds toward Indo-Africa . . . for long distances they are overthrust to the South against the Indo-African table-land" (Suess 1904–1909:I, 596).

Not only were the continents not oscillating, they were not even "the continents." Suess's Indo-Africa, the great tableland to the south of Eurasia, was composed of Africa, Arabia, Madagascar, and

the Indian Peninsula and was once continuous across the Indian
Ocean, the latter viewed as a totally and permanently subsided
segment of the previously continuous continent (Suess 1904–1909:I,
600). In Suess's theory the continents and oceans were not structurally differentiated, nor were they permanent: "In considering
the main features of the structure we must not allow ourselves to
be deceived by the hydrosphere (Suess 1904–1909:IV, 505).

In the conclusion of the first volume, Suess extended the analysis to other continental areas. The convex island "festoons" along
the coast of Asia suggested that "a special tectonic homology exists
between that fragment of an ancient table-land [i.e., the Indian
Ocean], and this part of the ocean [i.e., the Northwest Pacific]": the
Pacific Ocean was the subsident foreland of these Asian ranges.
The Caribbean, in this analysis, became an American Mediterranean, with the Gulf of Mexico as a subsident foreland. Suess was
determined that the theory of continental oscillations should be
defeated not in the isolated treatment of favored locales like the
Appalachians or the Alps, but *everywhere*. Every structural feature
of the crust had to find a consistent explanation that was incompatible with the oscillatory theory but in accord with the theory of
contraction, tangential thrusts, and the "foundering," or subsidence, of continental fragments.

The specification of "tectonic homologies" and the redivision
and combination of the continental masses that Suess suggested in
1885 were largely an elaboration of ideas he had published a decade earlier in *Die Entstehung der Alpen*. To the refutation of the
plutonic uplift of mountain ranges he had added an assault on the
quietism of uniformitarian geology. He had attacked the oscillatory
model of continental evolution by accounting for the same phenomena in terms of subsidences produced by secular contraction.

However, as he himself pointed out, the principal reason for
the success of the oscillatory model of continental evolution had
not to do with structural questions but with the successful explanation of sedimentary cycles and faunal successions in terms of the
transgressions and regressions of the seas as the continents fell and
rose again. Moreover, the lack of explanations for such marine
cycles was a conspicuous weakness of all previous contraction theories, which had tended to concentrate on inundation following
cataclysmic uplift of mountains or on catastrophic flooding of a
more vague and general kind. The initial volume of his own theory
seems particularly vulnerable in this regard and little improved by
the repudiation of a general catastrophe. Oceanic subsidences

deepened the ocean basins, and the foundering of continental fragments widened them, but in neither case would the oceans be likely
to transgress the continents. Yet everywhere the sedimentary
record gave evidence of repeated advance of the seas onto the
continental masses.

The question was of sufficient importance that Suess devoted
the entire second volume of *The Face of the Earth* (1888) to its
solution. The elemental outlines of the argument are familiar,
though the matter of the discussion has changed. Previously the
reader had been shown the world from a mountaintop and had
been asked to join a celestial visitor at even greater heights; now
"we have descended from the mountains and stand on the seashore"
(Suess 1904–1909:II, 1). At our feet we see the tide advance and
retreat, while behind us

> traces of an older strand may be clearly seen, standing high above
> the existing level of the sea. For mile after mile they may be
> followed at a constant height, undisturbed by the nature or struc-
> ture of the coast. . . .
>
> This is something very different from the crushing and over-
> thrusting we meet with in the mountains. . . . The phenomenon
> before us is of an altogether different nature, and if we recall the
> play of the tides with their rhythmic rise and fall in phases of half
> a day, then nature herself seems to suggest the question whether
> other forces may not exist capable of causing much more impor-
> tant oscillations and of much longer periods. [Suess 1904–1909:
> II, 2]

Once again, we are reminded that the earth may be measured by
man but not according to man, either in its size or the duration of
its workings. And this reminder leads, as in the first volume, to a
consideration of the history of opinions on the subject. Beginning
with Strabo, Suess undertook a history of theories of displacement
of shorelines.[2] Early modern theories like those of Benoît de Maillet
(1656–1738) stressed a fall in the level of the ocean by a secular
decrease in the volume of oceanic waters. In the early nineteenth
century, theories of the uplift of the land, exemplified by Playfair
and Buch, generally accounted for oceanic displacement. In the
middle of the nineteenth century various theories of gravitational

2. The extensive history of studies of oceanic transgressions that Suess presented in the first chapter of vol. II of *Das Antlitz der Erde* is a useful source, but
experience with the tendentiousness of such histories by Lyell, and Suess himself,
indicate that it should be used with caution.

displacement of oceanic waters were put forward, most notably by J. Adhémar in *Révolutions de la mer* (1842). Adhémar theorized that the present attitude of the earth's axis caused the South Pole to receive fewer hours of sunshine than the North Pole, which favored a secular accumulation of ice at the South Pole. The enlargement of the icecap displaced the center of gravity of the planet and had drawn the oceans away to the south. With the precession of the equinoxes, the situation would change, and by A.D. 11,748 a maximum of refrigeration and oceanic displacement would occur at the North Pole.

> In spite of . . . many . . . weak points, Adhémar's work was most stimulating in its influence as a serious attempt to explain, by a single and consistent theory, three great phenomena, namely the predominance of water in the southern hemisphere, the periodic return of glacial epochs, and the universality and constancy of oscillations of the strand-line. [Suess 1904–1909:II, 19]

Suess argued that this and all the successor theories of its type could be reduced to the proposition that "there exist great and continuous regions of changing level (of the seas), the distribution of which is related to the rotation axis of the globe according to some easily recognizable law." The only alternative to such theories would be a "return, in spite of all objections, to the theory of movements of the solid crust" (Suess 1904–1909:II, 19). While gravitationally produced asymmetric accumulation of the seas was acknowledged even by such men as Darwin and James Geikie to be great enough to affect climate, the theory of oceanic transgression as a consequence of continental oscillation had prevailed as an orthodoxy chiefly owing to the efforts of Lyell.

> Charles Lyell, whose long career produced such important results for our science, was always a keen opponent of the theory of elevation of volcanic mountains, the theory that is, of elevation craters, and an equally keen and influential champion of the theory of the secular oscillations of the continents. On the formation of mountain chains he never expressed himself with equal decision. . . .
>
> In spite of the attacks of certain physicists, directed chiefly at the absence of a closer definition of the stupendous force, which is said to elevate and depress great parts of the earth's surface, the elevation theory has maintained its position as the accepted doctrine up to the present day, especially among geologists who devote themselves to stratigraphy; from this doctrine the explanation

of transgressions and gaps in the series of formations is derived, just as it was sixty or seventy years ago. [Suess 1904–1909:II, 22–23]

The tide of opinion had been so strong that even Suess himself had been swayed into interpreting the uplift of Scandinavia as a "formation of a fold of great amplitude within the earth's crust," a view that he now wished to reconsider.[3] Having decided that the "folding of the mountain chains and the formation of horizontal strand-lines which run without interruption over mountain segments of the most diverse description, are two entirely different things" (Suess 1904–1909:II, 23–24), Suess urged theoretical caution. In *Die Entstehung der Alpen* he had applauded Prévost's suggestion that the discussion of mountains should be conducted in terms of displacement, since the phenomena could be explained by either uplift or subsidence. Now his oracle was George B. Greenough, but the message was the same.

> In 1834, at the annual meeting of the Geological Society of London, not long after the publication of the last volume of Lyell's "Principles of Geology," the president Greenough warned the members against too ready adherence to the theory of elevation which was then rapidly gaining recognition. . . . the term "elevation" was used in the most various senses, and of the uniform elevation of a whole continent it was almost impossible to form a conception; above all, a *terminology* was needed which should not involve any preconceived theory. [Suess 1904–1909: II, 24]

Suess proposed to adopt the suggestion and speak not of elevation of the land but "negative displacement of the strandline," and similarly, to drop the term *subsidence* and speak of "positive displacement of the strandline." Even then he found the number of variables so great and the data so uncertain that "little remains as the result of many years labor, but a conviction that many doctrines which in spite of the warnings of unprejudiced authorities have become accepted dogmas are erroneous"; he determined that his treatment of the subject should remain "mainly critical" (Suess 1904–1909:II, 24).

3. Suess attributed this opinion, which he had expressed in *Die Entstehung der Alpen*, to his reliance on "the teaching of revered masters." More likely he was influenced by the theory of crustal warping proposed by Dana in 1873. Suess first accepted and later repudiated several of Dana's ideas. See Suess 1911:101–102; and 1904–1909:IV, 627.

In ensuing chapters he followed out three lines of evidence—
the extent of ancient seas, the nature of sedimentary formations,
and the examination of existing coastlines, as they bore on the
various hypotheses adduced to explain them. This included a survey of the Atlantic and Pacific coasts and their contrasts and a
detailed treatment of the generality or localization of various marine incursions. As one might expect, however, his particular aim
was the demolition of the uniformitarian theory of continental oscillations as an adequate explanation of the history of marine
cycles.

The theory stated that the elevations and subsidences of the
continents had to occur in slow, continuous increments and that
these processes must operate in the *present*. Suess's most obvious
targets were those instances of ongoing uplift that were offered as
proof by the uniformitarians—particularly the phenomena associated with the Temple of Serapis at Puzzuoli, Italy, and those
associated with the displacement of the strand in the Gulf of Bothnia. He devoted an entire chapter of the volume to *each* instance.

> The borings of marine shells which present themselves in a
> broad horizontal band on the pillars of the temple of Serapis at
> Puzzuoli, are adduced by numerous observers, and in most of our
> textbooks, particularly in Lyell's "Principles of Geology," as a
> proof of the repeated elevation and subsidence of the land. [Suess
> 1904–1909:II, 368]

Lyell had even gone so far as to make an engraving of the temple
pillars for the frontispiece of *Principles of Geology* as a symbolic
representation of the entire uniformitarian position. In typical
Suessian style, the examination of the case began with an analysis
of manuscripts dating back to 105 B.C. and traced evident fluctuations in the strandline in historical time. Suess's conclusion was that
the level had changed so many times that the land would have had
to bob up and down like a cork. If the land had oscillated so many
times, he felt it unlikely that the hot springs behind the temple,
which had remained open for two millennia, would not have been
affected by dislocations. His conclusion, not surprisingly, was that
the peculiar location of the temple in a volcanic subsidence had led
to a number of marine incursions—it was the *sea* that had risen and
fallen (Suess 1904–1909:II, 387).

In the following chapter, Suess dealt with the less tractable
problem of the Gulf of Bothnia. There, scientific measurements
over the course of a century had documented the alteration in

level; Lyell, Buch, and others had carefully examined the site in person (Suess 1904–1909:II, 411). Suess's approach was to discuss the number of unconsidered variables in the situation—that the gulf was fed by more than 200 rivers with great seasonal variation in stream flow, that wind and current controlled the outlet of these waters into the Baltic, and so on—to show that any positive proof of an elevation of the land, rather than a fall of the water, was lacking. His own opinion was that the gulf was emptying out into the Baltic and that the water level was falling.

> It is a question of climatology and hydrostatics, not tectonic ge-
> ology. That which happens is *an emptying out of the water, not a rise
> of the land.* Even if we were willing to concede so extensive an
> elevation of the lithosphere we should still be unable to explain
> why the movement should be restricted to the region of the Swed-
> ish coast and of the outflowing Baltic current. . . . As regards *the
> general secular elevation of the Scandinavian peninsula, the source and
> origin of the theory of elevation, definite evidence is entirely wanting.*
> [Suess 1904–1909:II, 413–415]

Suess had done his best to repudiate the evidence of uplift and had recognized that as regards the Baltic coast of Sweden, his account was only as plausible as uplift, and no more. Thus, at the end, he fell back to the argument that uplift, even if present along a limited section of coast, could not be adduced as evidence for a general uplift of a continental segment; this was a much stronger argu- ment.

Suess also devoted a chapter to the demonstration that the entire Mediterranean area lacked any evidence of secular uplift in the historical period—*or* any evidence of secular subsidence. There was no oscillation of any kind in this region, whose great tectonic activity he had chronicled in the previous volume. Here, in the presence of numerous earthquakes and volcanoes, which were the superficial expression of the uplifting forces of the Lyellian scheme, there had been no perceptible shift in the strandline since the beginning of recorded history (Suess 1904–1909:II, 465).

In the concluding section of the volume, Suess outlined his own theory of marine cycles. The tectonic element was subsidence: "The crust of the earth gives way and falls in; the sea follows it" (Suess 1904–1909:II, 537); however, there were different subsi-dences to be considered. When a portion of the continental crust sank, the effects were local; when a portion of the oceanic floor sank, the general negative movement of the oceanic level was

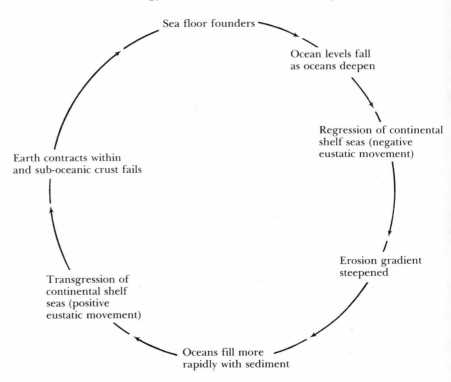

Sea floor founders

Ocean levels fall
as oceans deepen

Regression of continental
shelf seas (negative
eustatic movement)

Earth contracts within
and sub-oceanic crust fails

Erosion gradient
steepened

Transgression of
continental shelf
seas (positive
eustatic movement)

Oceans fill more
rapidly with sediment

Eduard Suess's eustatic theory of oceanic transgressions and regressions, driven by the contraction of the earth. Explaining oceanic transgressions was a difficult task for Suess, who postulated ocean basins widening and deepening through time.

worldwide. Suess termed such a general alteration of level of the strand a *eustatic* movement. "*The formation of the sea basins produces spasmodic eustatic negative movements*" (Suess 1904–1909:II, 538).

A consideration of the continental surfaces and their strata revealed that "vast areas have been subject, during periods of extraordinary duration, to marine transgressions, that is, positive movements which have been interrupted by negative phases" (Suess 1904–1909:II, 538). These positive movements had a different character, defined by the "extraordinarily slow rate" at which they have taken place and the extent of simultaneous transgressions on different continents. Deposits of the Middle Cretaceous transgression, he argued, could be seen in Brazil, western Canada, Europe, Egypt, Central Asia, Borneo, and Sakhalin. Suess concluded that "the formation of sediments causes a continuous, eustatic positive movement of the strand-line."

The thickness of the ancient sediments and the extensive denudation of the continents, amounting in many places to thousands of feet, show how great is the bulk of material which has been carried into the sea in the past. The oceanic regions are filled up slowly but without intermission, and their waters are in consequence gradually displaced; at the same time the transgression which thus results is facilitated by the progressive denudation of the land. [Suess 1904–1909:II, 543]

The effect of the alternation was compensatory in nature: each subsidence of the ocean floor stimulated river flow to a lowered ocean, increased the pace of erosion, and contributed to a greater eustatic positive movement as the sea filled in at a greater rate. Eventually, however, the subsidences would triumph.

Suess applied this interpretation of the sedimentary record to a refutation of the theory of secular oscillation. The differentiation of spasmodic negative movements bespoke a nonuniform process rather than a steady rise and fall of the land. Moreover, the universality of the positive and negative movements recorded in the sedimentary record suggested "changes . . . much too extensive and too uniform to have been caused by movements of the earth's crust." The fact that "it has been found possible to *employ the same terminology to distinguish the sedimentary formations in all parts of the world*" was evidence for the eustatic movements, and incompatible with the idea of the independent movement of individual continents (Suess 1904–1909:II, 540). A consideration of the time scale of the marine cycles argued that the uniformitarian theory was too uniform; consideration of the extent of inundations argued that it was not uniform enough. While Suess admitted that a number of oscillatory movements could be recorded that did not correspond to eustatic movements, he was adamant that the phenomena did not in any way conform with the theory of continental oscillation.

Although in regard to these questions so much still remains obscure and calls for further investigation, yet it is evident that of the hypotheses proposed to account for such movements, there is one series which cannot be reconciled with the facts, as these are already known to us.

Movements like these, which present themselves as oscillations, and extend around all coasts and under every latitude in complete independence of the structure of the continents, cannot possibly be explained by elevation or subsidence of the land. [Suess 1904–1909:II, 550]

Suess's eustatic theory was a great success and a major influence into the third decade of the twentieth century in the interpretation of the evolution of the continents, reaching a high point of popularity between 1928 and 1935 in the work of Viktor Baulig (Chorley 1963:953ff.). However, the most immediately influential part of the second volume, in the 1880s, was not Suess's general theory of marine cycles and his careful refutation of the theory of continental elevation and subsidence. Instead, it was Suess's chapter called "The Outlines of the Atlantic Ocean." While Suess had been busy preparing the first volume of *The Face of the Earth,* many striking "confirmations" of his theory of unidirectional thrusts had been published in France, Sweden, and Scotland. Under the pretext of surveying the Atlantic region, Suess devoted a fifth of the entire second volume to an updated summary of tectonic work on the European continent. For the first time a unified picture was given of the continent from Scandinavia to the Mediterranean, and the result was a new understanding of the evolution of Europe in terms of successive orogenic cycles.

> The extremely complex structure of Europe may be resolved into three principal folded chains, lying one behind the other, and all overfolded towards the north. The northernmost range, the Caledonian, is of pre-Devonian age. The second is composed of a western or Armorican segment, and an eastern or Variscan segment; they meet in syntaxis [*Scharrung*] in France, and the junction may be recognized even on the outer border, in the re-entrant angle which is formed near the Franco-Belgian frontier by the Coal measures overthrust to the north. This second chain is of pre-Permian age, but with posthumous foldings, the formation of which was continued at least as late as the tertiary era. The third chain, which is the most recent, includes the Pyrenees, the fragmentary arcs in the southeast of France, the Jura, the Alps, and the Carpathians. Each of these chains was dammed back in its course by the sunken ruins of the preceding chain. [Suess 1904–1909:II, 536]

The vast influence of this conception of the geological history of Europe in the decades that followed will become evident in the next chapter. Suess's ideas were most victorious in France, where the ground had been well prepared by Elie de Beaumont and Cuvier, but their effects in England and Scotland, Germany, Austria and Switzerland, and North America were scarcely less profound. Suess's work marked the inauguration of a "new global tectonics" and succeeded, within a few years of its appearance, in reversing the trend toward a more general acceptance of Lyellian uniformitarianism.

In its place, the contraction theory became dominant on the foundation of the structural evidence in the first two volumes of *The Face of the Earth*. Suess's detailed reconstructions of the mountains of the world and their connected history turned discussion of orogenic cycles from a consideration of their possibility to a determination of their extent and exact succession. Moreover, Suess reunited tectonics (the creation of structure) and orogeny (creation of mountains), which even contractionists like Dana had been willing to drop. Suess's work marked a resurgence of the trend going back to Werner to see the regularities of structure as the key to the history of the earth. Studies of sedimentary cycles, which under the aegis of the theory of continental oscillation had been the ultimate expression of geological synthesis, were reduced to the raw materials of a greater and more complex reconstruction. The history of the earth did not end with the elucidation of the sedimentary record; it began there. Suess bore out and extended Dana's dictum: earth history without structural evolution is earth history with the history of the earth left out.

Suess's work gave an entirely new meaning to the term *uniformitarianism*. While his history of the earth was composed of secular processes proceeding in the manner and at the rate specified by Lyell, it also included and was even dominated by elements almost exclusively identified in the previous fifty years with catastrophism. Suess had pared away at the extremes; he divested the catastrophists of cataclysms while preserving the effects: oceanic inundation of continents and "rapid" and occasional creation of mountain systems against the background of the secular plodding of the erosion cycle. At the other end of the spectrum, the uniformitarian account of the structure of the crust as the random expression of infinitesimal and only accidentally incremental processes was completely eliminated from serious consideration.

The final eradication of the extremes of the time scale of geological change abolished the uniformitarian-progressionist debate in its traditional form. But the uniformitarians and the progressionists had also argued over the details of the vertical movements of the crust—whether they were mostly down, mostly up, or a periodic alternation of the two movements. It was Suess who first insisted on attention to horizontal movements and translations of crustal segments, not in crustal shortening or lateral crushing of mobile zones *in situ* or in the slow lateral growth of continents at the borders, but the tremendous lateral dislocations evidenced in the great overthrust mountain belts to the north and south.

Suess also transformed the content of geological theory by challenging in the most forceful way the idea of continental permanence, arguing that the present arrangements were only transitory; the distant future would be as different as the distant past. He accentuated for the first time the proposition that the structural history of the earth was best understood as the fragmentation of giant paleocontinents: the intercontinental correlation of structures revealed a previous, undeniable continuity.

These changes in the terms of discussion of geological theory were so rapid and so all-encompassing that a generation later the traces of the revolution were almost completely obliterated. (The old uniformitarianism was so convincingly replaced that it all but disappeared.) But Suess effected a change in geological methods that was certainly as momentous as his substantive contributions to theory: Suess made "working by the book" respectable. Suess's synthesis was a monument to the extraordinary value of sifting and mastering the geological literature already accumulated, rather than rushing off to some crucial locale to gain eyewitness verification or to produce one more field report. In the enlarged scope of Suess's synthesis of geology *all* locales became crucial—there is no analogue in Suess's theory to the Temple of Serapis or the Baltic tidemarks. The spectacular results of his method of work put a new pressure on the literature of geology and on the abilities of geologists; they marked the end of the age in which geology was a popular science and the observations of any literate amateur were gladly welcomed by a geological survey or journal. The literature was now too important to be entrusted to the tyro, and geology became a truly professional activity.

Suess was not the first to see that geology must be a cooperative enterprise; Lyell had said so in *Principles of Geology*, and Elie de Beaumont had built his theory on the work of distant observers in locales he had never visited. But the geology of each of these men was fatally dominated by an idée fixe. In Lyell's work the principle of uniformitarianism was a creed; Elie de Beaumont's fatal passion for the réseau pentagonal hardly needs to be reiterated.

Suess genuinely cooperated. In an age and a science noted for the bitterness of its fights over priority of conception and discovery, Suess was candid about the source of his own successes in the work of his predecessors. He recognized, as did his contemporaries, that his genius was not in the spinning out of conjecture or the originality of ideas but in the working out of an integrated master theory and in the substantiation or refutation of the theoretical claims of others. The notion of paleocontinents was suggested by the paleontological work of Melchior Neumayr. The idea of zones of diverse displacement (by faulting) and the differentiation of folded and block-faulted mountains came from G. K. Gilbert. The ideas on subterranean lava reservoirs and maculae originated with Clarence King and Clarence Dutton. The idea of radial subsidence came from Dana and Elie de Beaumont, and the idea of the "rigifaction" of plateaus from Dana. Studer had turned Suess's attention to the dislocation of the north foreland of the Alps, Escher to the importance of overfolds. Suess built his theory of the Caledonian orogeny in Scandinavia on the work of A. E. Tornebohm, and he worked out the theory of successive orogenies in Europe with Bertrand, his own student. He was deeply indebted to the work of Albert Heim and Ferdinand von Richthofen and was inspired by De la Beche, Cuvier, Elie de Beaumont, and Werner.

Suess organized this material in a way that provided an original, stimulating, and even decisive picture of the face of the earth. He united observations with an original terminology—"syntaxis," "virgation," "posthumous folding," "eustatic movement," "Laurasia," "Gondwanaland." The sequence of orogenies he proposed has found a permanent place in the literature of geology. (Hallam 1973). His mode of presentation, which assumed a knowledge of locales and structures as great as his own, made it difficult for many geologists to follow the details of his arguments; but the very recursiveness, repetitiousness, and exhaustive listing of examples was ultimately a great strength. Many of the great paleogeographic syntheses that followed were modeled on his and appeared as maps, graphically representing, often powerfully, the results of syntheses of the literature as great as Suess's own. But these efforts were weakened by the fact that the documentation for the syntheses remained in the mind of the synthesizers, while the sources of Suess's reconstructions were chronicled and available for all to see.

The tendency of this evidence was everywhere to support generalizations that cut philosophically and substantively against the uniformitarian position and to pave the way for the demise of Lyellian geology in late-nineteenth-century Europe.

CHAPTER EIGHT

The Nappe Theory in the Alps: Tectonics over Physics, 1878–1903

T HAT Eduard Suess should have gained an immediate international following for a theory unlike any of its predecessors in a science notable for the acrimony of its controversies was the result of a complex accident. Suess was a great geologist and a man of profound learning, but so were Lyell, Elie de Beaumont, and a host of others. The success of Suess's theory of the earth was not that it was vastly better than anything that had preceded it either— it was not. Suess had one advantage over all of the great geologists who had been his teachers—he outlived them. In the 1870s nearly all of the great founders of geology passed away: Murchison and Haidinger in 1871, Escher von der Linth in 1872, Adam Sedgwick and Louis Agassiz in 1873, Elie de Beaumont in 1874, Lyell in 1875, Karl von Baer, George Poulett Scrope and Sainte-Claire Deville in 1876. These men, all nearly exact contemporaries, had controlled the major university chairs and headed the national geological surveys. Elie de Beaumont had exerted enormous influence as head of the Carte Géologique, and as professor of geology at both the Ecole des Mines and the Collège de France. Sainte-Claire Deville, who succeeded him and was a strong exponent of his theories, lived only a year after assuming his professorship. Escher had been professor at the Polytechnich Institute in Zurich. Murchison was the director of the Geological Survey of Britain, Haidinger the director of the Austrian Survey, Sedgwick the professor of geology in Cambridge. Lyell was preparing the twelfth edition of his *Principles of Geology* at the time of his death.

The deaths of these men were not merely events to be noted with the traditional tributes and eulogies awarded to retired greats —most of them died in harness. Positions of great power and influence were suddenly opened all over Europe, often to much younger men whose reputations had yet to be internationally established. Further, as is sometimes the case, the death of the masters allowed the pupils to publish works at variance with the views of their seniors, views which had been withheld during their lifetimes out of respect or caution.

The réseau pentagonal of Elie de Beaumont perished as soon as he and Sainte-Claire Deville were no longer around to hammer it into the heads of French geologists and mining engineers, and a somewhat belated Lyellianism began to flourish in France (De Launay 1913:88). In Switzerland, Escher von der Linth's successor, Albert Heim, published in 1878 an interpretation of the massive overfolding of sedimentary beds in the Glarus district of the Alps which was the beginning of his great reputation as an Alpine geologist. The appearance of Suess's work in 1875 was a part of this changing of the guard. Things moved somewhat more slowly in Great Britain, where Andrew C. Ramsay became director of the Geological Survey in 1871. His directorship, as E. B. Bailey acidly remarked, is best remembered for the career of Charles Lapworth, who was not a member of the survey (Bailey 1952:90). However, Archibald Geikie succeeded Ramsay in 1882, and Geikie's *Textbook of Geology*, published that same year, was "written with the declared intention of widening the international outlook of British geologists (Bailey 1952:100).

The passing of these great men coincided with another significant development: pressure for the formal organization of the international geological community. There was considerable agitation for a regularization of terminology and mapping technique among the various national surveys; it was feared that the outpouring of new research using incompatible nomenclatures for time and stratigraphical divisions would create separate and mutually incomprehensible literatures to the detriment of the progress of the science. As a result, the first International Geological Congress was held in Paris in 1878, concurrent with the great Paris Exposition. In 1881 a second triennial congress at Bologna advanced the attempt at systematization begun at Paris, with stricter definitions of stratigraphic designations like *system* and *series*. After 1883 there was more general agreement on the names of the periods of geological history, and by 1884 some progress was visible in the attempt to standardize the scales and color schemes of the geological maps of the various surveys (Moore 1941:185).

This last point was of particular importance. The 1870s and 1880s were the great age of the national surveys begun earlier in the century in France, England, Germany, Austria, Switzerland, and the United States and Canada. The last blanks in the maps of the European nations were being filled in, and great stretches of Asia, North America, and the rest of the world were being systematically mapped for the first time. As detailed as the work of Lyell and his contemporaries had been, geological exploration had been still so limited that no truly worldwide summary and interpretation of geology had been possible.[1] By 1880 such a synthesis was not only possible but the need was pressing. Geological exploration was generating mountains of unassimilated, undigested material that bore on the solution of questions debated since the beginning of the century. International and intercontinental correlation of geological formations and cataloguing of mountain types and of the structural elements of the continents had progressed to the point at which the available evidence had, for the first time, actually outrun theory.

It was into this receptive atmosphere that Suess had stepped with *The Face of the Earth* in 1883, and the enormous success of his work is due in some measure to the necessity for such a synthesis at the very time he introduced it. The core of the work was his interpretation of the Alps as a mountain system displaced *en bloc* against a resistant foreland, and the method of analysis was a correction and amplification of the methods followed by Elie de Beaumont and Studer. The extent of the movement and the forces implied were far greater than anything imagined by Suess's predecessors, and by 1883 the movements he postulated were even greater than those he had suggested in 1875 in *Die Entstehung der Alpen*. His interpretation of the work of others had led Suess to his radical departure from generally accepted ideas of mountain formation. The novelty of his interpretation and the extent of the documentation he employed to substantiate it made his work the center of attention throughout the 1880s and 1890s.

Suess's work received immediate support from Albert Heim (1849–1937), who had recently succeeded Arnold Escher von der Linth as professor at Zurich and had inherited all of his manu-

1. George Sarton's "La synthèse géologique de 1775 à 1918" (1919) contains a capsule history of the geological surveys and geological mapping of this period.

scripts and unpublished notes with the chair (Renevier 1879:131). Heim had studied the Alps for many years, especially the Glarus district of the Swiss Alps. The district contained the *kolosalle Uberschiebung*, the great overthrust of older onto younger sediments, which Escher had discovered in 1841 but had been subsequently reluctant to discuss in print. "No one would believe me," he had said; "they would put me into an asylum" (Bailey 1935:50). In 1878, Heim published *Untersuchungen über den Mechanismus der Gebirgsbildung* (Investigations of the mechanism of mountain building), based in part on Escher's field notes. The work was a general treatise on the mechanics of rock deformation and the allied (but not identical) problem of the mechanism of mountain formation; it included a detailed study of the Tödi-Windgällen Gruppe, the section of the Alps containing the famous overthrust (Heim 1878).

Heim agreed with Suess on the inadequacies of the hypothesis of plutonic uplift to explain the rise of the Alps. In arguing for the passivity of central massives, Suess had highlighted the relative scarcity of supposed centers of uplift (*Erhebungscentra*) along the crests of Alpine chains and had shown how they failed to account for the uniform lateral movement and bowlike flexure of the northern border of the Alps. He had also strengthened his argument by correcting stratigraphical errors in the work of his predecessors who, like Studer, had sometimes argued for the uplift of young strata by older plutonic rocks that had long ceased to be molten or possess any active uplifting power. Suess had demonstrated that igneous or plutonic rocks were often not only interbedded with but identically flexed and deformed with the strata above and *below* them, strata that the plutonic rocks were supposed to have uplifted.

Heim agreed with Suess in these interpretations of 1875, but he was so disappointed at losing priority of publication in ideas arrived at independently that he had nearly abandoned his researches. In the end he published anyway, hoping to influence geologists to extend to other sections of the Alps the thoroughness of his own interpretation of the Glarus district. His work was incomparably detailed, and Suess immediately incorporated it into the first volume of *The Face of the Earth* in 1883. Heim warmly approved Suess's research into the contributions of American geologists (Dana and others), which had till then been unavailable in Europe, and stated that by supporting Suess's ideas in detail, he hoped to overcome "the shocking fragmentation of the literature on questions of mechanical geology (Heim 1878:I, vi).

Heim accepted wholly Suess's belief that the Alps were a one-sided structure caused by a one-sided thrust of the Alpine system of mountains to the north. However, in spite of this general agreement, Heim averred that the general movement of a system did not uniformly dictate the direction of the overthrusts, nor was the mechanism of the overthrusts and of the general movement of the system the same (Heim 1878:I, 234). He interpreted the Glarus district as a great double fold (*Doppelfalte*). The eroded and interrupted sediments were, for Heim, the remnants of two anticlinal ridges that had been squeezed up and folded over toward each other and then overthrust. This interpretation of the structure involved overthrusts from both the north and the south and was therefore at variance with Suess's interpretation of a single directional thrust that had overfolded the chain uniformly to the north. However, Heim argued only that there were necessarily areas that had been bilaterally compressed and truly upthrust within the borders of the northward moving system; the Glarus was such an area (Heim 1878:I, 234).

Despite this departure from the uniformity of Suess's scheme, Heim, in response to local particularities that demonstrated a greater diversity of movement, supplemented Suess's general picture with detailed elaborations of a specific section and corroborated the notion that such a general movement was physically possible. When Escher had first proposed the lateral overthrust of the great sedimentary slices of the Glarus, his lifelong friend Studer had opposed him on the ground that no sheet of rock could maintain its integrity while being overthrust and would instead be crushed; whatever the appearances, no such overthrust could have taken place.[2] Suess had done nothing to refute the original criticism from the standpoint of mechanics but had only pointed out the strong likelihood, from stratigraphic evidence, that such a reversal of sequence and overthrust had indeed taken place. But whereas Suess emphasized the broad general sweep of the Alpine system and remained at the level of a general overview, Heim worked from the other extreme of perception. He microscopically investigated the crystalline rocks of the Glarus and developed the theory that under pressure, even brittle rock underwent plastic deformation without losing its hardness (Mathews 1927:141).[3] The

2. Suess's reminiscence of this disagreement appears in *The Face of the Earth* as a part of the preface, written by Suess in 1904. (Suess 1904–1909:I, iv.)

3. See also Renevier 1879 and de Margerie 1946. While it is almost redundant to say that a work of de Margerie has bibliographic utility, this article contains much more than a survey of Saussure, Studer, and Heim, and is an excellent starting point for investigation of controversies in Alpine geology.

plasticity of such rocks was isotropic, and it was latent: only when lateral forces were applied did the rock become mobile. Thus he separated the *conditions* of deformation from the process of actual mountain building. Heim's evidence for this contention included still recognizable fossil mollusca that had been laterally stretched two to ten times their lengths (Mathews 1927:141). Heim combined this mechanical theory of deformation in the solid state with a theory of chemical change under pressure that explained metamorphism (Renevier 1879:133). Turning his microscope to the central "eruptive" granites of the region, Heim demonstrated their passivity in the tectonic displacements by showing that they were compactly folded and recrystallized and that the deformation and folding was of Tertiary age, contemporaneous with the creation of the sedimentary folded structures (Heim 1878:I, 239). Heim's detailed explication of the microstructure of the rocks in question and his paleontologically based demonstration of the plastic deformations of Alpine strata went a long way toward removing the obvious objections to the theory of massive overthrusts raised by Studer and often repeated by others.

Heim also strengthened the case for massive overthrusts in his treatment of the scale and meaning of erosion within the Alpine system. Buch, who developed the theory of axial uplift in the form accepted by Studer, had declined to accept erosion as a major factor in creating relief—for Buch, everything was a consequence of plutonic upheaval in some form. An inheritance of this conviction was the theory, still defended in the 1870s, that the deep Alpine valleys were fissures in the crust and certainly not the result of the puny streams barely visible from the peaks they had supposedly created. Heim rejected this idea forcefully in the first volume of his work and went on to assert that fully one-half of the mass of the strata composing the Alps had already eroded away (Heim 1878:I, pt. 5). Erosion in the Alps had not acted as a sculptor's chisel but as a carpenter's plane, leveling whole sections down to bedrock (de Margerie 1946:civ). This too was important for Suess, especially for the theory of eustatic changes in sea level that he was already formulating. If oceanic transgressions onto the continents were to be attributed to a filling of the seas with erosion products, then erosion on this massive scale was certainly necessary. Heim, however, had his own reasons for postulating such tremendous erosion. The idea of massive overthrust not only required demonstration that the overthrust beds were able to withstand the pressures involved without being shattered and destroyed but also had to allow for reconstruction of the original position of the beds across great distances separated by yawning valleys. The aspect of the theory of plutonic uplift that saw the valleys as tectonic structures—as a splitting apart—reduced the magnitude of the overthrusts Heim had postulated to a simple and local inversion. Only if the beds were once continuous across great distances could the theory assume tectonic importance, and Heim's theory required massive erosion to lend conviction to the idea that distant structures with no intervening fragments to connect them were once continuous series of sediments now separated by the secular action of water and ice.

Heim also had his own views on the way in which the contraction of the earth had made mountains. If he rejected local plutonic uplift, he also rejected Suess's notion of peripheral subsidence—the Alps were not high, he said, because the lands bordering them had sunk low. The overall contraction of the earth was expressed not in sinking but in tangential *shortening* of the crust, producing lateral movement of mountain ranges. Arguing from the amount of "packing" of the Alpine folds, and estimating the folding of the portions removed by erosion, Heim asserted that the Alps represented a zone of crustal shortening of about 50 kilometers (Heim 1878:II, 211–215). The Alps were a squeezed-together remnant of a once flat-lying section of the crust, which had adjusted to a shrinking interior not by sinking into it (by some method or other) but by shrinking in its turn, the shrinkage expressed as tight and extreme folding of the sediments, with great deformations on the scale observed.

These generalizations were not established without controversy. The theory of Alpine uplift through the plutonic upwelling of great masses of granite was reiterated as late as 1876 by Eugène Viollet-le-Duc, and even the postulation of massive erosion as the cause of deep Alpine valleys was disputed—the question of where all the eroded material had gone remained unsolved as late as 1928 (Heritsch 1928:195).

But the greatest controversy was over Heim's interpretation of overturned strata in the Glarus region as the remnants of a gigantic double fold and overthrust. When Ernest Favre, a partisan with Heim and Suess of the passivity of the granite masses of the Alps, reviewed the year's work in Swiss geology in 1882, the entire first part of his account was given over to the Glarus controversy. Heim's detailed case study of the Glarus was the foundation of his entire theory of compression, overfolding, and erosion. Many of the debates generated by the publication of Heim's work in 1878 had

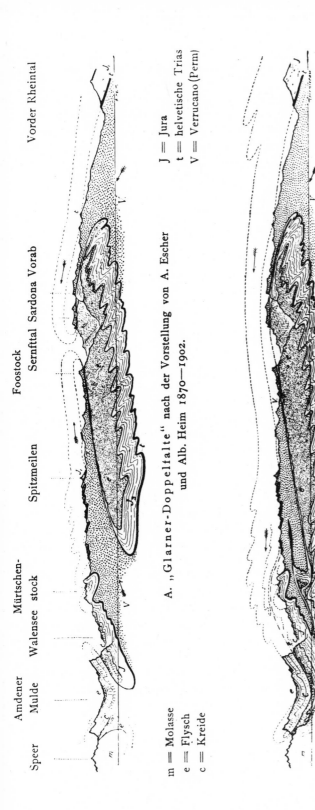

Speer　Amdener　Mürtschen-　Spitzmeilen　Foostock　Vorder Rheintal
　　　Mulde　Walensee　stock　　　　　Sernftal Sardona Vorab

A. „Glarner-Doppelfalte" nach der Vorstellung von A. Escher
und Alb. Heim 1870—1902.

B. „Glarner-Deckfalten" nach der Vorstellung von M. Bertrand 1883 und E. Sueß 1892, angenommen von Alb. Heim 1903.
Die Glarner-Deckfalten in schematischem Profil von N nach S nach älterer und neuerer Auffassung.

m = Molasse　　　　J = Jura
e = Flysch　　　　　t = helvetische Trias
c = Kreide　　　　　V = Verrucano (Perm)

Two views of the controversial Glarus district of the Alps. The legend reads: "The Glarus Decke-fold in schematic profile from north to south according to the older and newer conceptions. A. Glarus Doublefold according to the views of A. Escher and Alb. Heim between 1870 and 1902. B. The Glarus Decke-fold in the view of M. Bertrand in 1883 and E. Suess in 1892, accepted by Alb. Heim in 1903." (From Heim 1919–1921: III.)

been published, and in 1882 Heim led an excursion of sixteen senior geologists to the Glarus fold so that the matter could be argued on the spot. At that time, Heim picked up the influential support of the French Alpine geologist Charles Lory, and it was agreed by most of those present that the reversal of strata had occurred in the form of a great double fold. An Austrian participant, August Rothpletz (1853–1918), disagreed with the idea of a double fold. His objection is noteworthy because he hinted at an interpretation that later was the foundation of a theory of even greater dislocation: he argued that while the reversal of strata in one portion seemed to be an overthrust anticlinal fold, the northern portion seemed to him to be a *glissement*, a glideplane combined with a fault, and not a fold at all (Favre 1883:180). Yet none of the participants disagreed with the possibility of a lateral displacement on the scale that Heim had suggested, whether or not the actual structure was a double fold.[4] The principal problem lay in gaining assent to the mere possibility of such titanic movements in the horizontal plane rather than in establishing the mechanism of the movement or the actual direction of displacement; and this, Heim was able to accomplish.

While Heim's interpretation of the Glarus was weathering this early storm, one of Suess's Austrian colleagues challenged the notion of the unilateral northward movement of the Alps. A. Bittner argued in 1881 that the secondary chains of the Italian Alps and the southern side of the Eastern Alps showed folds and reversals of strata as great as those that Suess emphasized on the northern border. This proved to Bittner that the Alps were symmetrical and that the symmetry was clear evidence of a deforming action from the center of the chain outward (Bittner 1881:219). However, he was unable to account for the arcuate shape of the range to the north, a central point in Suess's analysis. Further, Heim's argument that overfolding in either direction could be expected locally regardless of the motion of the range as a whole served to blunt the force of the criticism. In spite of the differences between Suess and Heim on the specific mechanism of contraction that had produced the Alps, their interpretations reinforced one another against the prevailing view of bilateral symmetry and of upthrust at the center of the chains. Indeed, in the years that followed, the work of Heim and Suess, with the differences forgotten, was seen as a single theme in the discussion of Alpine structure.

4. Rothpletz was correct: the Glarus is not a double fold. Bertrand said so in 1884, Suess (privately to Heim) in 1892; Heim abandoned the interpretation himself after 1901.

The interest in the theory of massive overthrusts exhibited by European geologists was soon duplicated in Great Britain. Heim's work was introduced to British geologists in 1879 with a glowing review in *Geological Magazine* of his treatise on the mechanism of mountain formation. The reviewer's prediction that "the work is one which will certainly become classical" was a fine rhetorical flourish, but it was borne out by subsequent events (Renevier 1879: 135). However, it was a completely independent series of investigations simultaneously carried out in the Northwest Highlands of Scotland that provided evidence that gave international importance to the theory of overthrusts born in the Swiss and Austrian Alps. Charles Lapworth and Charles Callaway, working independently, had interpreted the Northwest Highlands as the site of large-scale overthrusts. This interpretation was at variance with the views of Murchison and Geikie, both of whom had headed the Geological Survey and whose views thus had "official" status (Lapworth 1883; Callaway 1883).

Lapworth's story deserves to be told in greater detail than is possible here. As a brilliant loner, unaffiliated with the survey, his work had met with hostility, and worse, indifference. While camped in the Highlands and documenting the overthrusts he had discovered, he had recurrent dreams of the great Moine thrust passing over his body, and he suffered, in the polite language of the time, a "nervous collapse." E. B. Bailey (1935:22) attributed it to the excitement of discovery; more likely it was caused by the prospect of another round of professional humiliation.

In this instance, however, Lapworth was vindicated. Archibald Geikie, as head of the survey, dispatched the Scottish geologists B. N. Peach and J. Horne (who were later to become famous with their remapping of the entire district in accordance with the new interpretation) to investigate the claims of Lapworth and of Callaway. Geikie joined Peach and Horne on the spot and agreed with their confirmation of an overthrust of at least ten miles. Geikie generally thought that geological investigation should proceed at a deliberate pace and opposed rapid publication; he was distressed with the struggles for priority of discovery, which he found inimical to the sound progress of science. In this case, however, he relented, recognizing the implications of the discovery. He personally arranged for rapid publication in *Nature* and insisted on full publicity of his own previous error. The effect of this open endorsement on the outside world was, as Bailey reports, "electrifying"—the new interpretation was received with acclamation and as a confirmation of the earlier reports of overthrusting in Switzerland, Belgium, and even Quebec (Peach and Horne 1884; Bailey 1952:110–112).

This detailed reinforcement of Heim's work, and indirectly of that of Suess, made the appearance of *The Face of the Earth* a momentous event. Structural analysis was now revealed as an important tool in establishing interregional correlation—as important as stratigraphy itself. Suess's work underlined the importance of a much more extensive search for patterns of horizontal movement in the crust than had previously been thought reasonable or even possible. The confirmation of widely scattered incidences of vast overthrust also served to validate Suess's method of work "by the book" against the overwhelming prejudice in favor of personal reobservation of any such claim as the only means of judging its adequacy.

Suess's theory of mountain formation and the associated idea of a unified structural plan for the entire continent of Europe gained ground rapidly in Switzerland and France. Typical of Suess's influence on younger geologists was the experience of Marcel Bertrand (1847–1907). That Bertrand should have adopted Suess's method of work as well as his tectonic ideas was a family accident. Bertrand had snored while Elie de Beaumont soared through his lectures at the Ecole des Mines, but he enjoyed fieldwork immensely. However, his father wished him to be close to home and used his influence to have his son transferred to the section of the Carte Géologique located in Paris. Here Bertrand was given a desk job that cut his fieldwork time to less than three months a year (Bailey 1935:143). To make up for this lack of field experience, Bertrand became a voracious reader in the geological literature of all the countries of Europe—English, Belgian, Austrian, German, Swiss, and French. After reading Suess's *Die Entstehung der Alpen*, Bertrand became increasingly devoted to the problems of Alpine geology (Termier 1908:163).

Bertrand was a strong partisan of Suess's notion of a one-sided movement of the Alps to the north. Having read Heim's interpretation of the Glarus, Bertrand saw, as had Rothpletz in 1882, that the northern (south-facing) overfold of Heim's double-fold theory could very easily be seen as a simple overthrust. While this latter idea magnified the amount of lateral displacement across the section, it accorded better with the overall plan proposed by Suess. All the thrusts that Bertrand had studied himself and all that he had seen in the work of other geologists were oriented in the same northward direction, right up to the border of the continent. He was particularly influenced by the Belgian geologist L. Gosselet and had seen his sections of the northerly overthrust of the Belgian coalfields (Gosselet 1879/1880:505), which Suess had employed to demonstrate the tendency of "folding to overtake subsidence" in the first volume of *The Face of the Earth* (Suess 1904–1909:I, 143). Bertrand insured the rapid passage of Suess's ideas into French. In 1884 he published an interpretation of the Glarus as a single over-thrust, comparing it with the Belgian coalfields as typical of the northward movement controlling the structural evolution of the continent (Bertrand 1884). Within a few years, however, Bertrand carried Suess's ideas in a new direction, which had profound consequences for geological thought. In 1887 he interpreted the history of Europe in terms of three great epochs of mountain building. The earliest and most northerly—the Caledonian system, extending from Scandinavia through Scotland and England to Ireland, was dated as pre-Devonian. The next system, which Bertrand called the Hercynian, cut across the older Caledonian in the south of Ireland, the south of England, and Wales and extended into Europe across Belgium and France to intersect the Alps. The Hercynian system was also a result of a Paleozoic orogeny, dated from the beginning of the Permian. The Alpine system delineated by Suess was the most recent epoch of mountain building in Europe; and this great orogeny of the Tertiary period determined the present outlines of the continent (Bertrand 1887; Suess 1887).

But it was not only the European continent, Bertrand thought, whose form was so determined. He maintained that the Caledonian and Hercynian systems were continuous across the bed of the Atlantic and found their extension in the Green Mountains and the Alleghenies of eastern North America. Adopting Dana's idea of continental growth by "marginal accretion," he argued that the successive orogenies of Europe were created by the annexation of geosynclinal belts around the Scandinavian core, and he proclaimed the identity of the Caledonian-Hercynian geosynclinal with the polygenetic Appalachian geosynclinal of Dana. In this fusion of Suess's and Dana's views, the geosynclinal formed the arc-shaped border of a former North Atlantic continent.

In 1888, in the second volume of *Das Antlitz der Erde*, Suess cautiously repeated Bertrand's postulation of mountains continuous across the bed of the Atlantic (Bertrand 1887:442). Such a postulation depended, of course, on his own work—it was Suess's idea that the Atlantic was a young ocean resulting from the subsidence of a continental fragment. Suess then suggested that there was also a structural continuity across the basin of the South Atlantic between the Mediterranean and the Gulf of Mexico. On the basis of faunal similarities, Neumayr had postulated a "Central Mediterranean Sea" that formerly connected the Old World and the New through the bed of a narrow transverse ocean between now-vanished Atlantic continents. Suess corroborated this on the basis of his structural plan and argued that the entire Eurasian girdle of mountains was born in this ancient sea, the "Tethys," which was linked structurally and faunally to the Americas as well. The Tethys was older than the Atlantic Ocean, and Suess implied that it was once the oceanic border of northern and southern Atlantic paleocontinental masses (Suess 1904–1909:II, 538).[5]

The idea of intercontinental correlation of geological structures, paralleling previous intercontinental correlation of paleontological assemblages, was an extremely important development in geological theory as an "independent" confirmation of the existence of great paleocontinents, whose fragments were separated by "younger" oceans. Bertrand's connection of the mountains of Europe and North America became a part of every major subsequent paleogeography; it was something that had to be either explained or explained away, as it connected the type ranges of Europe with the type range of North America (the Appalachians) on structural, sedimentary, and paleontological grounds.

Intercontinental correlation and international cooperation were the hallmarks of geology in the 1880s. The postulation of great overthrusts by Suess, Bertrand, and others had found confirmation in a number of locales, to which was added in 1888 A. E. Tornebohm's report of an overthrust in Scandinavia of more than 100 kilometers—an overthrust of an entire chain onto a foreland, as in the picture developed by Suess (Bailey 1935:24). That same year, the ongoing process of international standardization of terminology resulted in the publication of a bilingual glossary of tectonic terms, in French and German, constructed on historical principles. The authors were Albert Heim and the celebrated French geological bibliographer Emmanuel de Margerie (de Margerie and Heim 1888).

Beginning with simple folds and faults, de Margerie and Heim expounded tectonic features with simple diagrams and related them to type locales discussed in the geological literature. Gradually and methodically, the work proceeded to transverse horizontal displacements, and finally, to the complex relations of folds

5. See also Neumayr 1886.

and faults, horizontal and vertical movements actually to be found in three dimensions.

The glossary served to institutionalize the new scale of tectonic activity and to give it a vocabulary of equivalent French and German words. The work also gave a tendentious push toward acceptance of the massive thrusting of recumbent anticlinal ridges as the major form of overthrust displacement, particularly notable in the attempt by Heim and de Margerie to develop English equivalents for their terms. For instance, in the choice of the term *lying overfold* as an equivalent to *pli couché* and *liegende Falte* (de Margerie and Heim 1888:55), they tried to give reality to the European idea of the overthrusting of recumbent anticlines as a standard mode of geotectonic displacement. But Scandinavian, English, and Scottish geologists generally treated overthrusts as simple thrust planes rather than as remnants of complicated preexisting fold structures. When English-speaking geologists finally acknowledged the possibility of such structures on the large scale, they chose the term *recumbent fold*.

Of even more interest, considering the events of the next two decades in Europe, was the complete absence in the glossary of words that would soon dominate the study of Alpine tectonics—*nappe, Decke, Klippe, Fenster, zone des racines*, and the rest of the newborn terminology of the nappe theory, which would soon take the concept of overthrust tectonics to lengths undreamed of by the pioneers of the theory.

Like the uplift theory and the contraction theory, the nappe theory is a name for a group of ideas that were centrally predisposed to a particular kind of motion characteristic of large segments of the crust. According to the nappe theory, the Alps do not merely contain overthrusts, they *are* overthrusts. The entire Alpine system from France to the Carpathians (west to east) and from the Swiss plain to Italy (north to south) is, in the nappe theory, the remnant of a series of magnificent, superimposed thrust masses that moved over one another from south to north, displacing sections of the crust hundreds of square kilometers in extent. Each of the great thrust masses was a nappe, or in Germany a *Decke*, characterized by a distinct sedimentary assemblage traceable back to a root (*racine, Wurzel*) far to the south, from which it had been "squeezed out" by the contraction of the earth.

The easiest way to imagine the process is to conceive of the Alpine strata, before dislocation, as a complexly patterned tablecloth (*nappe*) on the surface of a highly polished table. If you should place your hand flat on the table and push forward, the cloth will begin to rise into folds. Push more and the folds will flop over (forward) and the rearmost fold will progressively override those before it, producing a stack of folds. Take a pair of scissors and cut away at the stack from various angles, removing whole sections of the pile of folds. Having done so, push the pile again so that the segments become jumbled against each other. Then invite a friend into the room and ask him to reconstruct the pattern of the original tablecloth. With great difficulty, using the elements of the pattern he can see, your friend painstakingly examines the pile and mentally recreates the continuities of the pattern across the gaps you have cut out, following the lines of the pattern to separate the original folds, now jumbled, into their original relations. Then he mentally unfolds the tablecloth onto a flat surface and conjectures the remainder of the pattern across the gaps that still remain. This process, at a level of analogy commensurate with the use of a shriveled apple to represent the contracting earth, is the reconstruction of the Alps in the terms of the nappe theory. The tablecloth with its complex pattern is the stratigraphic record; your scissors, erosion; your friend, the Alpine tectonist, developing a picture of the original extent and relations of the sequences in an act of clever imaginative reconstruction—working from pattern and continuity alone. Like the true Alpinist he does not compute a coefficient of frictional drag for the table or inquire as to the plasticity of the tablecloth; he draws lines, assuming that the pattern in each part of the cloth is unique and that the various sections will fit into the whole in only one way. With the aid of these simplifying assumptions he is able to reconstruct the whole original flat surface.

As in the case of the contraction theory it was not the brilliance of the original conception but the detail with which it was worked out that made it significant. Nor did it begin from such a complete and schematic view. Like Suess's theory of the general movement of the Alpine chain, it began in the northern foreland of the Alpine system, the portion that Studer had warned long before would compel geologists to "enlarge our view of its cause" as its dislocations were more carefully studied. From there, the conception grew to encompass the entire Alpine system, as the estimates of the dislocation were revised ever upward.

It is extremely difficult to assign priority of discovery to the nappe theory. Certainly Suess's general conception of a northward movement of the Alpine system, Heim's theory of great overfolded and overthrust anticlines, and Bertrand's reinterpretation of the

Glarus double fold as a single thrust moving to the north all played an important role. One must also include the observation of Studer (1825) that the Molasse, the sedimentary material that borders the Swiss Alps to the north, contains fragments that cannot be matched with any of the foremost part of the Alps from which they might be expected to have been eroded. To this list must be added the observations of Arnold Escher in 1839 and 1853 and of F. J. Kaufmann in 1876 that a number of peaks known collectively as the Mythen were composed of a sedimentary assemblage foreign to the locale, occurring, inexplicably, in the "sea" of younger Tertiary sediments known as the flysch (Bailey 1935:73–89).

For years geologists had struggled with the anomalous facts of this region—the Pre-Alps, as it came to be known. Into the early 1890s the most popular theory was that a chain to the north of the Swiss plain, now eroded away, had provided the anomalous sediments observed by Studer in the Molasse. This imaginary "Vindelician chain" was preserved only in the Mythen, it was argued, and it was these older outcrops that poked through the Tertiary flysch sediments and produced the picture of older beds above younger; thus was the uniformitarian doctrine of superposition defended.[6] More simply, the Mythen and other peaks, composed of Triassic, Permian, and Carboniferous sediments, had their contiguous base buried in the Tertiary flysch that surrounded them on every side; this flysch could be correlated with the Alps to the south, from which it had eroded away.

However, in 1893, the Swiss geologist Hans Schardt discovered a serious flaw in this reasoning. While it was pleasant to be able to see the Mythen and the other peaks "poking through" younger sediments, Schardt had discovered that sections of these "buried stocks" were overthrust, to the north and the south, *onto* the flysch (Schardt 1893a:707; 1893b:570). Escher had noted the reversal in 1853, and Favre had explained it in 1867 as local faulting, but neither had noted that the pattern was repeated in *every single one* of the outcrops or peaks above the flysch. This provided a real anomaly for Schardt. It was all very well to see the Mythen as the exposed remnants of a very old chain largely buried by the flysch sediments that engulfed their bases. But by what reasoning could one explain a renewed tectonic activity that could cause each of these isolated peaks, independently, to overthrust the newer sedi-

6. It would be impossibly cumbersome to replicate here the stratigraphic reasoning on which all this was based. See Bailey 1935:84–85; for a longer account, see Heim 1919–1922:II, 3–72.

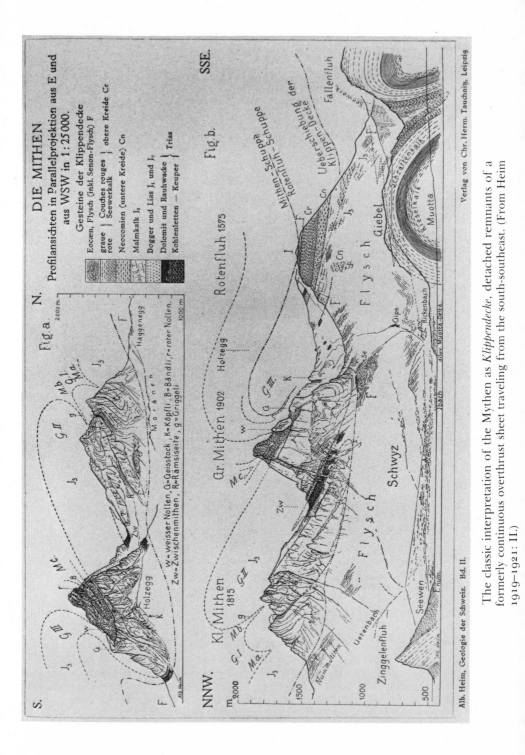

Verlag von Chr. Herm. Tauchnitz, Leipzig

The classic interpretation of the Mythen as *Klippendecke*, detached remnants of a formerly continuous overthrust sheet traveling from the south-southeast. (From Heim 1919–1921: II.)

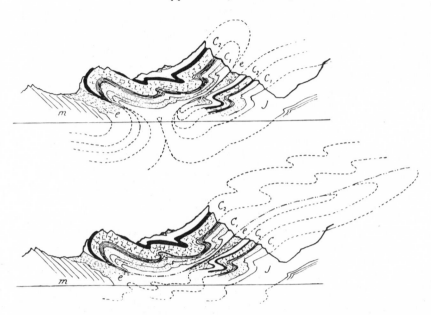

Two possible interpretations of an "exotic": *top*, as a *Pilzfalte* ("mushroom fold"); *bottom*, as a *Klippe*, or outlying fragment of a former *Decke* ("nappe"). Albert Heim's 1921 diagram shows how readily an exposure of this type can lend itself to both interpretations (dotted lines). (From Heim 1919–1921: III.)

ments to the north and south? Another prominent Alpinist, Maurice Lugeon (1870–1953) tried an interpretation based on the older ideas of the vanished (and buried) Vindelician chain to the north; he advanced the idea that the Mythen peaks were remnants of "mushroom folds." This tectonic oddity, then already well documented, is exactly what it sounds like: an emergent anticlinal fold develops a "cap" that flops over in *two* directions, producing a mushroom-shaped structure and an apparent overthrust onto the flysch.

However, Schardt knew of other outcrops east of the Rhone in which the same facies appear again and again (i.e., the same sedimentary sequence). The idea of an entire family of mushroom folds was to Schardt the reductio ad absurdum of the idea that the Mythen and the other protruding masses were the remnants of the vanished Vindelician mountains; such structures are so rare that a regular repetition, involving everywhere the same beds, was too much to contemplate (Bailey 1935:85).

There was yet another interpretation which might "save the phenomena." By 1892, other geologists had noticed the same anomaly as Schardt and were prepared to see the Mythen and the other "islands" in the flysch as masses of the vanished Vindelician chain overthrust to the south onto the flysch; the distance the masses would have traveled was thus short, and their vanished "roots" could be postulated as lying below the Molasse of the Swiss plain.

Yet everything written by Suess, Heim, and Bertrand in the 1880s had corroborated the idea of a uniform northward movement of the Alpine system against an older foreland. Moreover, from the work of Bertrand (1884), Schardt knew of certain features in the Mythen and the Pre-Alpine outcrops that were stratigraphically compatible with structures in the Rhaetian Alps (to the east and south) as far south as Provence. All this led him to postulate the idea in 1893 that the Mythen and similarly composed structures were the only remnants of a once-composite thrust sheet that had slid northward off the main mass of the Alps, gliding by the force of gravity far from the area in the south where it had been laid down and later uplifted (Schardt 1893a).

Thus arose the difficulty in assigning priority to the nappe theory. While later geologists traced their interpretations of the nappe overthrusts in the Alps to Schardt's postulation concerning the Pre-Alps, it is clear that Schardt relied generally on Heim and Suess and on the evidence provided by Escher, Studer, Gustav Steinmann (1856–1929), Carl Schmidt (1862–1923), Maurice Bertrand, Ulrich Stutz, Alphonse Favre, and especially Maurice Lugeon, in bringing forth the idea of the Pre-Alps as a remnant of a thrust sheet transposed to the north. Moreover, Bertrand helped Schardt to publish and even to prepare his findings (Bailey 1935:86).

However, it is clear that Schardt and Lugeon were responsible for the rapid development of the theory of massive overthrust sheets—nappes—in the 1890s. In 1896 Lugeon, at first critical of Schardt's idea of a composite thrust from the south, adopted the idea and published a paper detailing the same phenomena south of Lake Geneva to the banks of the River Arve that Schardt had covered north of Lake Geneva to Lake Thun. Lugeon's paper extended the idea of a composite thrust mass to the entire front of the Pre-Alps, which idea Schardt reviewed and again extended in 1898 (Lugeon 1896; Schardt 1898).

It should be noted that as Schardt and Lugeon strove to outdo one another in the explication of the phenomena of the thrusts of the Pre-Alpine foreland, their impulses were never, at any time, anything but conservative. Neither foresaw, in the 1890s, the extent to which they or other geologists would carry the postulation of great Alpine thrusts, and their instincts were always to explain the phenomena with a minimum of translatory motion of the masses involved. The nappe theory, in spite of its hypertrophic overlays of recumbent anticlinal remnants, transported over ever greater distances to maintain continuity with some autochthonous mass of rock, was devised in the service of the basic uniformitarian postulate of superposition—if older rocks were found atop younger rocks, some dislocation must have occurred. The notion of tectonic movements of tens and even hundreds of kilometers was dedicated to the preservation of the stratigraphic and paleontological sequences that had been established over the previous century and that was geology's only firm foundation.

The postulation of the Vindelician mountains to the north of the Swiss plain might have been a solution to the origin of the Molasse and the Pre-Alpine remanent blocks from the standpoint of stratigraphy and paleontology alone. However, the overall structural plan of northward movement delineated by Suess and Bertrand would then have to be abandoned and with it the entire tectonic frame of analysis that gave the two-dimensional picture of strikes and flexures any meaning or explanation. If the Vindelician mountains existed at all, the only evidence of their existence was either buried, detrital, or in the slightly less grandiose displacement of the exotic blocks of the Mythen and Pre-Alps from the north rather than the south. Contrary evidence suggested that successive orogenic episodes had, since late Paleozoic time, always resulted in transposition to the north of the site of deposition. Given these criteria, the postulation of a massive overthrust sheet from the south was the most logical and *conservative* explanation.

While Schardt and Lugeon vied with each other for the correct overall interpretation of the Pre-Alps, the French geologist Pierre Termier (1859–1930) had discovered similar overthrust structures far to the south in the Italian Alps (Termier 1899). Since the publication of the first volumes of *The Face of the Earth*, the early converts to the theories of Suess had striven to support his picture of the Alpine system by reexamining component mountain systems for evidence of overthrusts of the type postulated. In 1887 Bertrand had tried to establish a connection (suggested by Suess) between the Alps and the Pyrenees through an interpretation of the Beausset region of Provence as homologous with both the Glarus thrusts and the Franco-Belgian coal basin. Between 1888 and 1892 Emmanuel de Margerie extended this interpretation into the Pyrenees, showing that overthrusts were at least as important and perhaps more important than vertical faults in the creation of the structures of the chain (Fleure 1955:187).

Pierre Termier had been a student and close collaborator of Bertrand since 1888, and in 1891 had established his professional reputation by mapping successive stages of regional metamorphism in the Alps. Termier demonstrated that the crystalline schists of the great axis of the Alps were, as Charles Lory long before had suspected, not plutonic rocks but metamorphosed Mesozoic sediments. By 1894 Termier had accepted the Suess-Bertrand theory of successive orogenic episodes and had demonstrated the superposition of Alpine on Hercynian folds (Haller 1976:283–284). Having absorbed the ideas of Suess, Bertrand, and more recently of Lugeon and Schardt, Termier was working at the southern border of the Alpine system to see to what extent the theory of overthrusts could be confirmed there and a homogeneous picture of the Alps thus produced.

While Termier was at work in the south, Lugeon published a synthesis of his tectonic work since 1894 and proposed an interpretation of the Western Alps as a series of superimposed thrust sheets, now deeply eroded but still distinguishable, extending the entire breadth of the chain (Lugeon 1901a). Moreover, Lugeon had been won over to the idea that these thrusts were the upper halves of great anticlinal folds, drawn out across the entire upper surface of the Alpine chains. This postulation was based on the correlation of the stratigraphic succession of the Pre-Alpine thrusts with sequences far to the rear of the Alps, from which they had been cut off by erosion (Lugeon 1901b:45ff.). These sequences he called the zone of roots (zone des racines).

Lugeon's paper, more than 100 pages long, and its description of superimposed nappes, traveling from south to north, explained the puzzling picture of Alpine stratigraphy that had misled and confused observers as far back as H. B. de Saussure in the 1770s. In trying to sort out the jumbled outcrops of folded rock it had largely been assumed that the rocks were folded in situ—that they were indigenous to the locales where they were found. This, argued Lugeon, was a fundamental mistake. While it was true that the Alps had traveled north as a system, the rocks which had once been in the rear of the chain were now the farthest forward and had completely overthrust the entire Alpine system.

The impact of Lugeon's interpretation was heightened by its publication with an open letter from Albert Heim endorsing the picture Lugeon had given and repudiating the idea of an anticlinal double fold at Glarus. It was now clear to Heim that what he had thought was a double fold of autochthonous sediments was actually the remnant of a once continuous fold of much greater dimensions, far removed from its source in the south of the system.

Lugeon was at this time only thirty-one years old. Heim, at fifty-four, was the dean of Swiss geology and the head of the Swiss Geological Commission, and thus his endorsement of the theory was powerful. Leon Collet, whose *Structure of the Alps* (1928) is the most complete exposition of the nappe theory in English, was a student at the time, and recalled: "This letter, published in Lugeon's memoir, had great repercussions. . . . The adhesion of the famous geologist to the Nappe theory made a profound impression on those who had, till then, doubted the value of the theory. . . . I shall never forget the feelings of my comrades and myself on reading Heim's open letter, for we were all unbelievers" (Collet 1938: cxii).

Later that year, 1901, the long-awaited third volume of *The Face of the Earth* appeared, thirteen years after the second. In the opening pages Suess acknowledged the success of the nappe theory in the Alps while refraining from committing himself to it. "The study of folded ranges by means of transverse sections has recently been marked by great progress. . . . even the boldest hypotheses, advanced only a few years ago, as to the extent of the horizontal movement, stop far short of the totality" (Suess 1904–1909:III, 2).

And yet a major step in the progression of the theory was still to come. Termier had followed Lugeon's paper by only a few months with one of his own, cementing the relations of the French and Italian Alps as nappe structures. In 1903, the International Geological Congress met in Vienna, and the nappists used the excursions that regularly followed such congresses to put forward their views. Lugeon pointed out that the Carpathians could be best explained as nappes (Bailey 1935:129). But Termier had bigger game in mind. Ever since he had heard Lugeon's general interpretation of the Western Alps Termier had considered the possibility that the nappe theory could effect a total stratigraphic unification of the Alpine system from east to west as well as from north to south. Having accompanied Professor F. Becke's excursion to the Zillertal on the west side of the Hohe Tauern, Termier there announced, grandly, that the exposures of the Hohe Tauern were the key to the unification of the Eastern and Western Alps: the reason that no continuity had ever been established between the east and west is that the Eastern Alps are made up of a series of nappes that *completely override* the nappes of the Western Alps.

The Eastern Alpine nappes had traveled from south to north

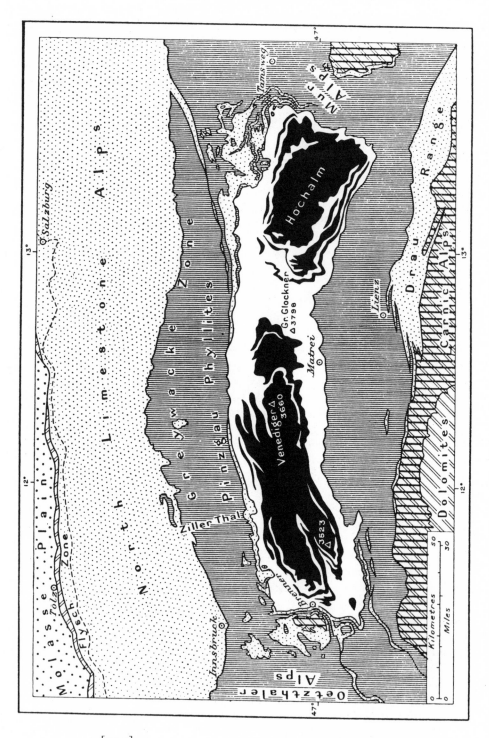

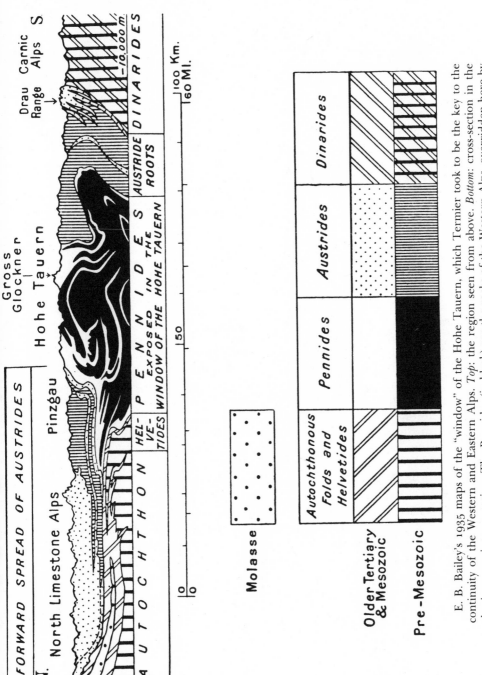

E. B. Bailey's 1935 maps of the "window" of the Hohe Tauern, which Termier took to be the key to the continuity of the Western and Eastern Alps. *Top:* the region seen from above. *Bottom:* cross-section in the classic nappe interpretation. The Pennides (in black) are the rocks of the Western Alps, overridden here by thrust sheets now completely eroded away farther to the West. (From Bailey 1935.)

on the back of the nappe sequence revealed by Lugeon in the Western Alps. The Hohe Tauern exposure was a "window": looking down into it, one saw at the lowest level of sediment the formations that were at the summit of the Western Alps—the Pennide nappes (Termier 1903:739, 742, 753). The reason why the Alps had never been successfully correlated along the strike of the system was that the sediments exposed in most of the eastern portion were *superimposed* on the sediments of the Western Alps. The difference between the two sections of the system was not structural but their relative state of preservation: in the West, the "Eastern Alps" were already completely eroded away. Termier showed that the "roots" of the East Alpine nappes were to the south of the Hohe Tauern—everything to the north of it was made up of nappes, some of which had traveled as far as 120 kilometers (Termier 1903:744).

Termier's views were the foundation of the "classic" theory of the Alps, which persisted in its general outline into the 1920s (Heritsch 1928:11); the beautiful simplicity and comprehensiveness of the conception was only reluctantly surrendered (Rutten 1969: 191–192). The basic theory of nappes had been superimposed onto the old "foreland-fold zone-backland" theory of Suess. The nappists divided Suess's fold zone into three parallel segments—the external, axial, and internal zones of the Alpine system—each of which had given rise to a series of nappes. In Termier's history of the system, following the Variscan (pre-Permian) orogeny, a geosynclinal depression formed to the south, roughly parallel to the older Variscan range, in which sedimentation was nearly continuous from the Carboniferous to the Eocene. These sediments, trapped between resistant foreland and backland, were folded and refolded.

Following Suess, Termier took the southern border of this geosyncline to be the region between the south border of the Eastern Alps (Austrides) and the Dinaric Alps (Dinarides) (Termier 1903: 754–756). In this borderland, the lateral pressure caused the development of a great "fanfold"—the northern segments of which had advanced as nappes. At some point, facilitated if not determined by a general subsidence of the Alpine region, the northernmost portion of the Dinaric backland (the internal fold zone) began to move uniformly and rapidly to the north. In the form of a solid sheet, this segment of the backland formed a *traîneau écraseur* ("crushing sledge"). This huge composite segment of the internal zone overrode the axial and external zones of the Alps, folding and breaking up as it surmounted the nappes already in its path. In the Eastern Alps of the present day, the only exposures of the external zone appear as flysch at the northern border of the system—all of the external nappes, including that of the Glarus, remain buried beneath the great thrusts from the internal zone. The nappes of the axial zone of the Alps, which appear as the Pennine Alps, or Pennides, in the west, are the portions revealed in the "window" of the Hohe Tauern, and are otherwise completely buried in the east beneath the great mass of the Austrides, the internal zone, and the Dinaride backland nappes (Termier 1903:754–756).

Although Termier carried off the prize of the long-sought "Alpine synthesis" with first publication of the overthrust of the Western Alps by the Eastern, priority, as in the case of the Pre-Alps, is difficult to determine, and ultimately pointless. From the first, the nappe theory was the product of a group of investigators in constant communication, close collaboration, and often keen competition. The intellectual lineages are simple. Suess trained Bertrand, Bertrand trained Termier. Lugeon was a student of both Heim and of E. Renevier, who had publicized Heim's work. Termier, Lugeon, Bertrand, and Schardt were in constant touch with each others' work. These men were succeeded in the interpretation of the Alps by geologists they trained—Emile Argand, Elie Gagnebin, Rudolf Staub—who continued to develop the theory into the 1920s. The "new tectonics" in the Alps was rapidly established by a concerted campaign that used every available forum to bring together exponents of the theory of overthrusts. In 1892, Bertrand, de Margerie, and Richthofen made a tour of the Northwest Highlands of Scotland accompanied by Peach and W. J. Sollas (Bailey 1952: 134). In 1893, Lugeon led an excursion of the Swiss Geological Society to the Pre-Alps. In 1894 the meeting of the International Geological Congress in Switzerland provided an opportunity for Heim, Schardt, and Lugeon to make known their views on overthrusts and to conduct excursions to locales like the Glarus and the Pre-Alps.

Throughout the 1890s the pressure was maintained, with a growing sense of excitement over successive discoveries of new and larger overthrusts in the Alps. And yet, after the first full approximation by Termier in 1903, the development of this picture of the Alps followed the same conservatism that Schardt and Lugeon had shown, aimed at the development of a stratigraphically credible version of the history of the Alpine system within Suess's postulation of the general northward movement of the Alps and the two-

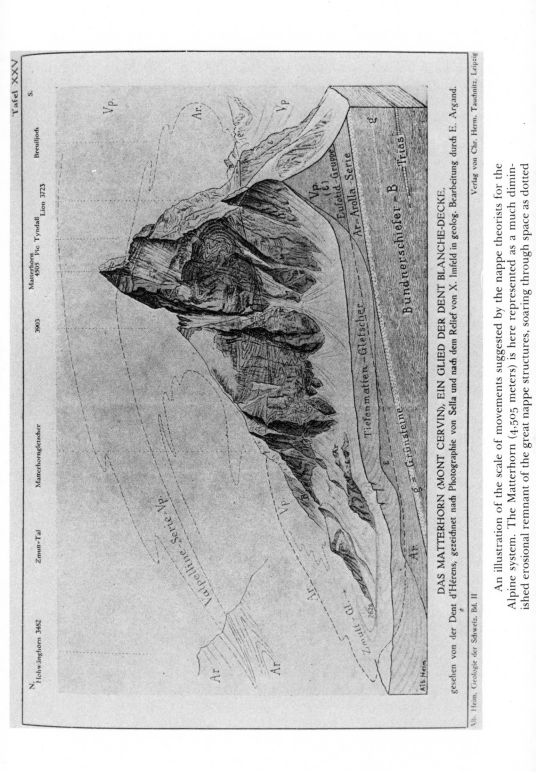

N. Hohwänghorn 3482   Zmutt-Tal   Matterhorngletscher   3903   Matterhorn 4505 Pic Tyndall Lion 3723   Breuiljoch   S.

DAS MATTERHORN (MONT CERVIN), EIN GLIED DER DENT BLANCHE-DECKE.

gesehen von der Dent d'Hérens, gezeichnet nach Photographie von Sella und nach dem Relief von X. Imfeld in geolog. Bearbeitung durch E. Argand.

Alb. Heim, Geologie der Schweiz, Bd. II   Verlag von Chr. Herm. Taudnitz, Leipzig

An illustration of the scale of movements suggested by the nappe theorists for the Alpine system. The Matterhorn (4,505 meters) is here represented as a much diminished erosional remnant of the great nappe structures, soaring through space as dotted dimensional framework of trend-line analysis. Schardt and Lugeon had both originally seen the Pre-Alps as gravity-driven thrust sheets and not necessarily great fold structures. As the nappe theory developed, Bertrand, Schardt, Lugeon, and Termier had tried to see the Swiss nappes as *écailles arrachées* ("torn-off flakes") rather than *plis couchés* ("recumbent overfolds") in order to minimize the amount of dislocation and lateral movement (Termier 1903:743). When maintenance of stratigraphical continuity and explanation of reversed sequences dictated the general picture of great squeezed-out folds that had traveled *at least* 100 kilometers, they avoided the idea of great folds arching through space and the vaulting of an already established mountain system by postulating a general subsidence of the Alpine region at the time of the great thrusts. Termier, in order to render the idea of the great crushing sledge even marginally plausible, was impelled to assert that the entire process had occurred at or near sea level on a gentle slope and only later had the Alpine region been irregularly and unequally uplifted.

The nappists had been able to generate powerful support. Heim came over in 1901 with his open letter accompanying Lugeon's memoir. Suess, in 1905, gave his seal of approval to the nappe interpretation of the Alps after personally confirming it in a study of the Eastern Alpine windows. Termier, as "field general" of the French geologists in the dispute over the interpretation of the Eastern Alps, was generally recognized as the disciple of Bertrand, who was already past president of the Société Géologique and a member of the Academie des Sciences. By 1911 Termier was director of the Carte Géologique, by 1914 inspector general of mines; he held office as president of the Société Géologique and as vice-president of the Académie des Sciences. By 1911 Termier was director of the theory was rapidly and completely established as the official orthodoxy in Switzerland, Austria, and France. With the resignation of Geikie as head of the Geological Survey of Great Britain and the accession of Horne to the direction of the Scottish section in 1901, the lid was lifted on unofficial publication by members of the survey, and increased attention to the problem of overthrusts was evident in Great Britain almost immediately (Bailey 1952:164).[7] The mood everywhere was one of buoyant optimism: it was generally

7. Horne had found Geikie's restriction extremely galling. Since the lag of publication was enormous, careers had been damaged and many memoirs never appeared at all. He had sworn that if he were ever in power, his subordinates would never suffer similar treatment. He kept his word. Not till Sir John Flett (1869–1947) assumed control in 1911 did the "old order" of prior censorship return.

felt that, as Termier had said in 1903, the basic lines had been laid down and the day in which the picture would be complete was only a few years off. By 1903 the time of grand announcements and astonishing confirmations had passed, and Alpine geologists turned increasingly "within" to elaborate the grand structural interpretation they had devised.

CHAPTER NINE

New Orogenic Theory and Intercontinental Correlation

ALTHOUGH the work of Suess and the inspiration provided by the first volumes of *The Face of the Earth* left an unmistakable imprint on the work of the nappe theorists, they were not the only members of the geological community thus affected. Nor did all the nappists pay exclusive attention to the elaboration of the three-dimensional picture of the Alpine system and its evolution that was first suggested by Suess's picture of the overthrust of the Alps onto their northern foreland. By the time of the appearance of the second volume of *Das Antlitz der Erde*, Suess himself had turned his attention from the particularities of the Alpine system to the more general geological history of the European continent and its relationship to the other continents and oceanic basins. When, in 1901, he applauded the success of the Alpine geologists, he nevertheless insisted on a broader view as the main goal of geology.

There was a time when every single anticline of the Jura was regarded as an independent axis of elevation; then it became clear that such a collection of parallel anticlines must have a common origin; next it was seen that there is a certain dependence between the Alps and the Jura; finally the influence of the obstacle presented by the Black Forest was recognized, and it became evident that the Alps and the Jura were only part of the southernmost, innermost, and most recent of three crescentic systems of folds which have arisen one after another across Central Europe since the close of the Silurian epoch. Thus with our increasing knowledge we are led to the conception of units of continually ascending order, and the several anticlines of the Jura now appear to us as parts of an organic whole.

To continue this method of synthesis, to group the folded ranges together in natural units of still more comprehensive character, and to explain by means of a single expression as large a part as possible of the terrestrial folding—such is the task that now awaits the geologist. The plan of the *trend-lines*, written by nature on the face of the earth—this is what he has to determine. [Suess 1904–1909:III, 3]

Suess's aim was consistent from the first volume to the last—to follow out the trend lines. His explanation of the existence of the trends reduced directly to the phenomena attendant on contraction and to the relationship of contraction to marine cycles. As early as 1875 he expressed a certain diffidence concerning the dynamic theory of the geosynclinal that had been postulated by Dana, and he buried his own speculations on the finer mechanisms of the crust in the long discursive summaries of the relations of the structural units. The nappe theorists followed Suess in this caution over the mechanical specifics of the creation of structure; accepting the great thrust from the south into Europe, and working out its consequences in the tracing of the nappes back to their roots, they were nearly indifferent to the mechanical problems raised by the extent of the lateral movements they proposed. They were equally indifferent to the depositional history of the sequences whose deformation they studied, except as it related to correlation of strata over long distances. Once again, the same argument that Dana put forth concerning James Hall's theory of mountains—that it was the origin of mountains with the origin of mountains left out—might be laid at the door of the nappists—but with a twist. While Hall had concentrated on the origin of mountains as the deposition of thick sequences of sediment and had ignored the structural evolution of the complex subsequently, the nappists stood at the opposite extreme: they studied the structural transformation of sedimentary sequences with little explicit attention to the way in which they had been formed. This complacence was due to their confidence in Suess's model and to the nature of the controversy in which they found themselves at the turn of the twentieth century. As M. G. Rutten later commented, "One gets the impression that geologists at the time were not so strongly interested in what lay behind these structures, being fully occupied in proving their existence" (Rutten 1969:180). In this latter task the nappists were extremely effective. But their work remained narrowly focused and ignored the larger implications of Suess's work—the relationships of continents and oceans. The origin of continents and oceans and the global interconnections of mountain ranges were the subject of a number of new hypotheses in the last part of the nineteenth century, either derived from or parallel to Suess's work. The general turn toward the history of the continents (through the analysis of former structural continuity) was a victory for Suess and his notion of successive orogenic epochs over the whole face of the globe. The reality of such epochs accepted, many European theorists turned to possible explanations for the periodicity and for the geographical distribution of the great structural features of the globe.

Marcel Bertrand, in particular, displayed a theoretical ebullience in the 1880s and early 1890s with various attempts to modify and extend Suess's structural plan. Most Alpinists agreed with Suess's substantive conclusions about the trends and movement of the Alpine system and concentrated on working out the consequences of that picture, but Bertrand was also influenced greatly by Suess's method of work—the refinement of the existing literature into coherent hypotheses without undertaking extensive new fieldwork. While it has been urged that the nappe theory was an "armchair theory," this is not the case—Schardt, Lugeon, and Termier spent most of their careers in the field tracing the convolutions of the nappes and searching out the root zones in the southern border of the system, from which the great overthrust fragments of the north and central portions had originated (Rutten 1969:179).

Bertrand, restricted to Paris most of the year, developed from the literature several rather abstract versions of the evolution of Europe, deduced from the pattern of folding in the rocks of the successive orogenic episodes. His earliest attempt in this direction was his postulation of a general northward overthrust in the Glarus region, in place of Heim's and Escher's double fold, and the relation of that overthrust to Gosselet's picture of the Belgian coalfields.

Then, in 1887, he combined the notion of successive orogenies in Europe—Caledonian, Amorican-Variscan, and Alpine—with Dana's notion of continental accretion in borderland geosynclinals, "marginal accretion"; it was here that Bertrand boldly postulated the continuity of the major orogenic systems across the Atlantic, pointing to a former unification of the two continents in a subsided Atlantic paleocontinent.

In 1892 he proposed an even more extreme version of this theory, strongly reminiscent of the *géometrie de précision* sought by

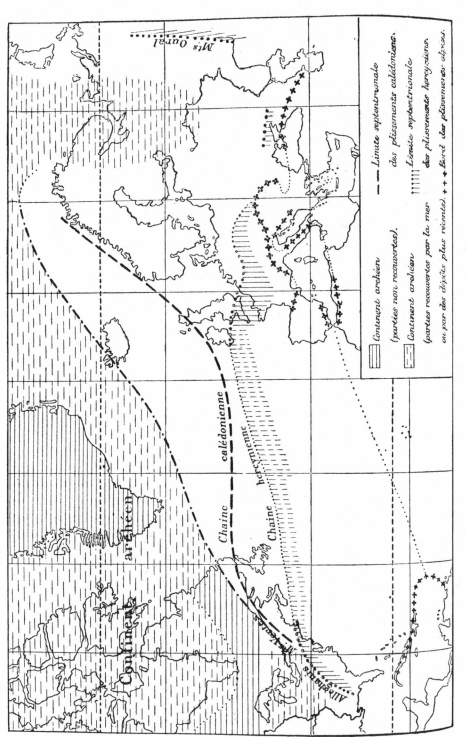

Marcel Bertrand's 1887 map attempting to establish continuity of the major European and North American ranges across the Atlantic basin. The Caledonians are supposedly continued in the Green Mountains of Vermont, the Hercynian ranges in the Alleghenies, and the Alpine ranges, represented as small crosses, reappear in Santo Domingo, Cuba, and the northern coast of South America. Note that their historical succession is the process

Elie de Beaumont in the trend of mountain ranges. While the successive ranges of Europe demonstrated a transverse folding directed to the north, there was also evidence of another pattern superimposed upon these successive overthrusts: a series of regular undulations in the long axis of the ranges from east to west, rising and falling at regular intervals, at right angles to the northward thrusting; he was alerted to this by observations of folding in the Paris basin (Haug 1907–1911:I, 520).

In 1892, Bertrand, by now completely convinced of the solidity of Suess's work as the foundation for the "general comparative orography" that Suess had hoped would someday be developed, grandly announced that the time had come for the development of *laws* of the formation of mountains in certain patterns across the face of the earth. Working from maps of trend lines and from the postulation of series of right-angle folds, he proposed a *réseau orthogonal*, in which the dislocation of the earth's crust was seen to occur preferentially in two directions. "The earth is progressively deformed along a network of orthogonal curves, the first set circumpolar and the second converging toward the polar regions" (Bertrand 1892:404). The corrugation of the earth's crust proceeded continuously and, in the same locales, always along parallels of latitude and meridians of longitude, oriented to the magnetic pole of the earth.

This represented the "outer limits" of theory at the time, and even Bertrand quickly backed away. Between 1891 and 1894 he suffered a series of embarrassments and setbacks that sobered him considerably. In 1891, Heim published again on the Glarus question and refrained from even mentioning Bertrand's name in the bibliography of pertinent work. Collet later wrote that this was in fact a snub directed at the interpretation of a geologist who had not visited the locale in question (Collet 1938:cxii). In 1892, Bertrand visited the Northwest Highlands of Scotland expecting to find evidence of broken overfolds (of the Alpine type); instead he found the clean-cut thrusts faithfully represented by Peach and Horne (Bailey 1935:86). A visit to the Glarus in 1893 caused him to doubt his own interpretation of the structure. In 1894, Heim prepared a paper on the Glarus for the guidebook of the International Geological Congress and again failed to mention Bertrand in the bibliography, though Heim was already, at Suess's urging, beginning to accept the idea of a single thrust (Bailey 1935:52, 88).[1]

1. Only much later, in *Geologie der Schweiz*, did Heim credit Bertrand with the correct interpretation in conjunction with Suess. See Heim 1919–1921:II, 12. Bertrand had by then been dead thirteen years.

Although much of Bertrand's work between 1892 and 1894 was governed by the idea of the orthogonal network, he soon realized that he was, as he told Termier, pursuing a chimera (Bailey 1935: 157). In fact, most of his work derived from the literature had come under attack from others and had been shaken by his own fieldwork. Thereafter he devoted himself to work based more on direct observation and less on the literature.

Aspects of these theoretical excursions nevertheless managed to find a home in the work of the Alsatian geologist Emile Haug (1861–1927), who developed a unique variant theory of intercontinental correlations that involved a return to the theory of continental uplift.

Haug was one of the few French geologists not swept into the Suessian fold in the 1890s. An experienced stratigrapher and widely read in the literature of geology, he was a gadfly to the nappists in the formative years of the theory and opposed Schardt's interpretation of the Pre-Alps in 1899. Very much later, in 1925, he posed a comprehensive challenge to the assumptions of the nappist Emile Argand on the reading of the stratigraphical evidence of the Alps, which had been offered as proof of Argand's theory of "embryotectonics" (Haug 1925). Although Haug confirmed Bertrand and Termier's interpretations of overthrusts in the Alps, with his own fieldwork, and even claimed priority of discovery over Termier, his interpretation of the meaning of these structures in the history of the European continent differed markedly from that of Suess and the nappists, which by that time had nearly achieved normative status (Carozzi 1972:168; Termier 1903: 766).

Haug's work is interesting because it occupies a curious middle ground between the interpretations of Suess and the picture of continental evolution developed by North American geologists about the same time. Haug was obviously influenced by Suess, but he was equally influenced by the idea of the geosyncline as an important concept in the unraveling of the geological history of Europe, a view that Suess explicitly repudiated (Suess 1904–1909:IV, 627). And yet Haug's interpretation of geosynclines, their structure and depth, their relation to continental masses and their history, is a sharp contrast to the outlines of the theory proposed by Dana.

The parts of Bertrand's work that attracted Haug were the application of the geosynclinal theory to European geology in Bertrand's 1887 paper on successive orogenies and the pattern of orthogonal folding sketched out in his theory of 1892. Haug was critical of the general conclusions—marginal accretion and the parallelism of "*grands circles de comparaison*"—but found the basic plan of successive geosynclinal depositions and·deformation and the pattern of right-angle folding suggestive of the true nature of the orogenic succession in Europe and around the world (Haug 1900:664).

In 1900 Haug published a long paper in the bulletin of the Société Géologique outlining his own history of Europe: "Les géosynclinaux et les aires continentales: Contribution à l'étude des transgressions et des régressions marines" (Geosynclines and continental platforms: a contribution to the study of marine transgressions and regressions). It followed the contemporary trend toward tectonic analysis as a key to earth history, but with important reservations. Haug agreed that dramatic results could be achieved by a study of comparative geology, in which mountain chains, continents, and oceans were viewed as organisms whose structure obeyed determinate laws. Suess, using continental fragments and islands to reconstruct and evoke the whole history of the past of these organic units, had successfully followed the lead of Cuvier, who had reconstructed the skeletons of biological organisms from fragmentary bones (Haug 1900:617). Further, Haug agreed with Suess that the study of shorelines was of only secondary importance in understanding the laws that govern marine cycles: the geological history of the continental masses and the oceanic abysses was profoundly more important in such reconstructions. This blending of tectonic analysis of continental features like mountain chains with the correlative history of marine cycles mirrored Suess's twofold concern with the significant portions of the geological history of the globe— the production, and later the deformation, of sedimentary assemblages on the continental platforms.

Yet Haug's reservations about Suess's results were great. Significantly, he rejected the eustatic theory in favor of the epeirogenic (continent elevating) oscillation. Suess's rejection of the uplift of Scandinavia left Haug completely unconvinced—the land *had* risen and continued to rise. And yet, worldwide swings in the level of the sea were amply confirmed, and their explanation was a pressing matter if the theory of continental elevations was to be maintained—Suess had raised the strong objection to continental uplift as an explanation of marine cycles by pointing out the unlikely occurrence of simultaneous oscillation of widely scattered continental units.

Haug accepted the idea that there had been extensive paleocontinents that included not only the present continents but substantial portions of the oceanic abysses. He agreed with Suess and Neumayr that once a narrow sea, the Tethys, had linked the Mediterranean and the Caribbean and had separated North and South Atlantic paleocontinents. He resurrected Rogers's old argument that the Appalachian sediments had been fed into the region from an older continent to the southeast, in the present Atlantic region. He supported on stratigraphic grounds Bertrand's idea of the continuation of transverse ranges linking Europe and North America across the Atlantic depression, stressing the continuity of the Belemnite chalks of New Jersey and Europe (Haug 1900:636). He postulated a transatlantic continuity between South America and South Africa based on paleontological similarities of Devonian age observed in the Cape Colony and Brazil. Further, he accepted the principles of compressional orogenesis, the foundering of continental fragments as the cause of oceanic depressions, orogenetic epochs in the history of the continents, and synchronous marine transgressions over the face of the earth. And yet, while accepting all these elements of Suess's theory he rejected the universal power of contraction, the notion of unidirectional thrust in orogeny, and upheld the doctrine of synchronous continental oscillations, against Suess's one-cause, one-tendency model of crustal evolution based on subsidence alone.

Haug's solution was the elimination of the great oceanic abysses throughout most of geological history. The present juxtaposition of continental platforms and profound depressions, resulting from the *morcellement des continents* ("breaking up of the platforms") was, he thought, a unique and geologically recent event. The characteristic picture of the crust throughout geological history was of giant paleocontinents separated by relatively narrow and deep geosynclinal depressions, poised in a dynamic balance. The earth was once laced by a system of narrow, interconnected seas like the single Tethys of Suess, when the continents were very much larger than they are today. The mechanism of transgression and regression was conceived simply: when the continents were uplifted, the oceans collected in the deep geosynclinal depressions; when the continents sank, the seas overflowed the borders of the deep troughs and created the broad epicontinental seas of the typical transgressive epoch. The huge size of the continents and complete circumscription by geosynclinal troughs explained the great, often global extent of the simultaneous transgressions and regressions.

Although Haug's geosyncline had some of the same functional

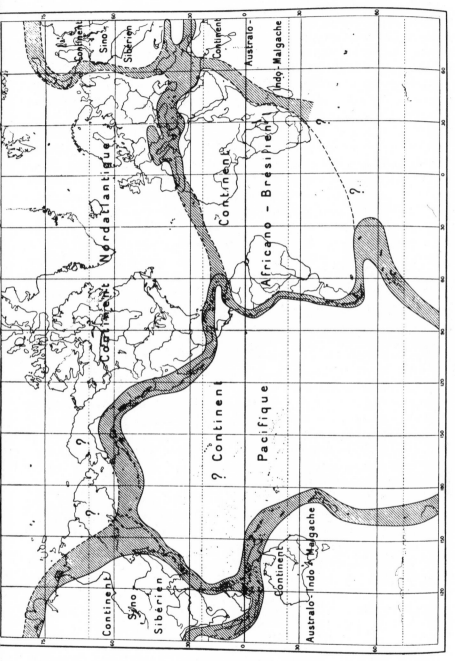

Emile Haug's (1900) view of the disposition of the major paleocontinents and their geosynclinal borders, superimposed on a Mercator projection of the present continents and oceans. (From Haug 1907–1911: I.)

significance as Dana's—particularly the idea that mountain ranges
were derived from geosynclinal sediments—he substantially altered
the picture of its structure and evolution. To begin with, Americans always assumed that geosynclines were formed successively in
the oceanward advance of the continental edges, and Haug's picture of very large, narrowly separated continents could not accommodate this idea, which he felt was quite mistaken (Haug 1900:
630). Haug's geosynclines were long-lived mobile zones between
relatively stable continental masses. This idea led Haug to postulate
a huge foundered continent over the entire bed of the Pacific Ocean
and to see the great circumpacific mountain arc from Borneo to
Tierra del Fuego as the remnants of a circumpacific geosyncline,
continuously active through the Devonian period.

Not only these mountains but all the great folded zones of the
earth were the result of activity in narrow geosynclinal mobile belts
between stable platforms. The Himalayas were such between Asia
and India, the Alps between Europe and Africa, the Carpathians
between the Russian platform and the Dinaric Alps, the Pyrenees
between the Massif Central and the Iberian Meseta, and so on.
These *intracontinental* geosynclines were significantly deeper than
the geosynclinals of Dana. Both Hall and Dana had argued for
gradual subsidence of a trough along the border of a continent:
Hall because he mistakenly believed the crust to be subsiding under
the accumulating weight of sediments, Dana because the nature of
the sediments seemed to indicate that they had been deposited in
shallow water; however great the total thickness, the process must
have approximated a "foot for foot" subsidence.

Again, Haug disagreed. To begin with, his geosynclines were
preexisting downfolds of the crust, from 100 to 900 meters deep.
In them, sedimentation was not neritic (shallow) as in the theory of
Hall and Dana, nor was it pelagic or abyssal, as Suess and Neumayr
seemed to suggest; it was bathyal—of intermediate depth. Haug
argued that both paleontology and lithology supported a bathyal
depth for geosynclinal sedimentation.

Applied to the history of the European continent, the scheme
worked as follows. After a more ancient orogenic phase, the continent was eroded away and transgressed by the seas. Subsequently,
a geosynclinal depression formed parallel to the axis of the preexisting orogenic phase and began to fill with grossly detritic sediments. At the beginning of every geosyncline there was formed a
median geanticline, dividing the geosyncline into two secondary
geosynclines. As the geosyncline filled, the waters were forced out

onto the continents to fill parallel depressions of lesser extent: a slow, continuous transgression.

The folding of sediments in geosynclines had always been a subject of debate. Hunt and Hall, following the traditional, even Huttonian picture, had argued for deformation in the period of infilling, followed by uniform continental uplift and erosion. Dana had argued for a "catastrophe" of plication by lateral pressure toward the end of the life span of each geosynclinal, producing folding, thrusting, and some elevation, followed by a more general continental upwarp. Haug's theory of folding was all-inclusive. In the first stages of sedimentation, "folds" appeared as deposits conformed to anterior folding, which was the consequence of previous orogenic movements and the presence of the median geanticline. Folding occurred after deposition, too, and here Haug declined to speculate on the exact cause, while pointing out that the parallelism was nearly perfect between the axis of major fold systems and the axis of the geosyncline. One can easily read the suggestion of some form of lateral pressure, in spite of Haug's reserve. At some point the geosyncline experienced an orogeny—an epoch of folding, faulting, partial foundering, and compression (*tassement*) (Haug 1900:675).

Haug then postulated an additional source of folding *after* the orogeny. Here he made his own contribution to the question of whether orogenic or epeirogenic movements were the source of mountain ranges: they were not necessarily alternatives, he said, but could be subsequent phases of a general movement. Bertrand and others had noted that entire bundles of folds (*faisceaux des plis*) could have longitudinal undulations that exhibited regular patterns. Bertrand had theorized in 1892 that this was caused by a general tendency for folding to take place concurrently in two directions at right angles to each other—an orthogonal network. This was only an appearance, said Haug. The phenomena did not result from two concurrent *orogenies* but from the succession and perhaps even the simultaneous occurrence of orogenic *and* epeirogenic movements. In short, the *axial* folds were either depositional accommodations or were orogenically produced, and the transcurrent folds were the response of the folded structures of the geosyncline to continental oscillations.

To develop this idea, Haug originated the concepts *aire de surélévation* and *aire d'ennoyage*, which may be translated as "regions of tectonic culmination" and "regions of tectonic depression." He envisaged these as small-scale replications, on the continental plat-

forms, of the pattern of the continent and its surrounding geosyncline (Haug 1900:665ff.). Areas of tectonic culmination were encompassed by peripheral depressions, areas of tectonic depression
by peripheral ridges. These alternations, caused by actual uplift of
continental segments, explained all the complications of relief and
right-angle folding.

Like Dana, Haug imagined that mountain systems were not
single concordant series but the superposition of several geosynclines with discontinuities between them. Although the axes of successive geosynclines were usually coincident, they need not be so.
Sometimes the new geosyncline cut across preexisting folds, as in
the Alps. This was to be expected—structural complications would
naturally increase through time as a result of the interplay of the
geosynclinal cycle and the development of regions of tectonic culmination and tectonic depression. Taken altogether, the general
parallelism of complex mountain fold systems remained constant
through time.

Everything depended, in the end, on movements of land. The
up-and-down movement of the continents alternated the amount
of pressure against the sediments in the peripheral geosynclines.
When the continents rose, the areas of their basis contracted and
the geosynclines were enlarged, corresponding to the phase of
general regression of the seas and the infilling of the geosynclines.
When the continents sank, their base area increased and caused an
increase in the pressure (lateral) on the geosynclinal sediments;
simultaneously, the seas transgressed the lowered continents. The
primary folding of geosynclinal sediments was synchronous with
periods of oceanic transgression; the orogenic phases corresponded
to oceanic regressions.

While strict in the correlation of the marine cycles with periods
of deformation and orogeny, Haug was deliberately vague about
the cause of the cycles. He argued that one could locate the dynamic motor in the movements of continents, with the geosynclines
as passive recipients of the attendant forces, but that it was equally
reasonable to imagine the reverse: a contraction and deepening of
a geosyncline could release pressure on the bases of the continents
and cause them to subside and even to founder; an expansion of
the geosyncline could compress the continental bases and force
them up (Haug 1900:708).

Haug's overall scheme had some notable strengths and obvious
weaknesses, but as an antagonist to the theory of secular contraction it was quite formidable. Its strengths were that it explained,

credibly, the geological history of Europe in its manifold complexity on the basis of a mechanism different from that of Suess, employing the geosyncline and the oscillation of continental movements. Between 1888 and 1900 Suess's theory had been uniquely successful in explaining the origin of mountains, the history of the continents, *and* the history of marine cycles with arguments based on contraction alone. Haug showed that the same synthesis could be achieved by a modification of the theory of continental oscillation, once the "true" history of the oceanic basins was revealed: their recent origin. He argued against the "too imaginative paleogeographies" of recent European geology—the theory of the orthogonale network, the overschematism of the nappe theory—while incorporating all the structures that they explained under the aegis of respectable concepts: epeirogenic movement and geosynclinal evolution of complex (polygenetic) mountain systems (Haug 1900: 617).

Moreover, Haug could explain how the Himalayas could be composed of marine sediments, though trapped between two paleocontinents, without the ad hoc postulation of a unique transverse Tethys, elsewhere unexampled in the history of the oceans. In Haug's theory, it was one of many intracontinental troughs that had given birth to every major mountain range on earth. He could also give an account of the mountains of the west coast of South America, which Suess admitted did not follow the regular pattern of "folding overtaking the foreland." The complexities of Alpine stratigraphy and transverse folding and the repetition of orogeny in Europe could be dynamically interpreted without the inexplicable triple recurrence of mountain epochs widely separated in time, successive from north to south.

On the other hand, the weaknesses of this theory could not be easily attacked by the followers of Suess. Suess himself had implied the foundering of the Atlantic and of at least the North Pacific according to the exigencies of the foreland-backland scheme. Although Haug's theory of late, unique, oceanic depressions through continental foundering was ad hoc and not dynamically related to the rest of his theory, it was no more catastrophic than the theory of the pure contractionists, who postulated the same phenomena responsive to the same mechanism. As for his giant Pacific paleocontinent, once the principle of continental foundering was established, who was competent to limit its extent? Haug still joined with Suess in the opinion that the Pacific was the oldest of the oceans and that the Atlantic was the youngest, and he followed Suess in many other structural deductions—opposing him principally on the final rather than the efficient cause of structure.

Whatever the differences and similarities between the theories of Suess and Haug, both involved acceptance of intercontinental connection in the past and the severance of that connection through the foundering of massive continental fragments as the oceanic basins were created and enlarged. Along with the theory of horizontal overthrust within the continents, the doctrine of the *imper*manence of oceans and continents stands as one of the great generalizations of late-nineteenth-century European geology.

CHAPTER TEN

The Decline of the Contraction Theory: The Challenge from Geophysics, 1870–1909

B ETWEEN 1880 and 1910, under the aegis of the contraction theory, a rough general picture of the evolution of the earth's crust was rapidly sketched out and extended over the face of the earth. The technique of trend-line analysis of major structural features, the theory of marine cycles based on the periodic foundering of ocean bottoms and the eustatic theory of sea levels, the demonstration of the intense deformation and extensive lateral translation of great crustal segments, the intercontinental correlation of periods of mountain formation, the localization of deformation in long, repeatedly active mobile belts loosely identified as geosynclines, all served to organize the phenomena of geology into a satisfying whole and to encourage a buoyant optimism that the day of geological synthesis had finally arrived.

This general optimism was challenged by geodesists and geophysicists who maintained that the central mechanism from which so much had been deduced—the theory of the secular contraction of the earth—was incapable of producing the extent of deformations observed. The very success of the structural geologists in proving that the crust had been ridged up and overthrust and their measurements of the amounts of overthrust and crustal shortening proved the undoing of the central mechanism—it was soon demonstrated that no estimate of the cooling and contraction of the earth, however generous, could account for the amount of folding already observed. Whatever role remained for contraction theory, it was not a universal explanation for the deformation of the crust.

The challenge emerged slowly and obliquely; the opponents of the contraction theory were at first concerned with the determination of the earth's figure (shape) and density, a problem that led them indirectly into the arena of tectonic theory. In the 1880s and 1890s the internal debates of the community of geophysicists and geodesists (largely separate from the world of geology) increasingly involved geological data, and the geological implications of the several geophysical hypotheses came to be discussed. One of the first results of these discussions was the general opinion that whatever hypothesis of cooling and redistribution of mass in the interior of the earth was true, the contraction theory in the form developed by geologists from structural evidence and Laplace's cosmology was almost certainly false.

The older form of the contraction theory, and geological theory in general throughout most of the nineteenth century, had only the scantiest mathematical and physical foundation. Every major theory developed since the time of Hutton and Werner had contained some postulation about the interior of the earth and its behavior, but the supporting evidence for any particular hypothesis was woefully insufficient. It was known that the earth was hot and dense within; the eruption of volcanoes was incontrovertible proof of great internal heat in some portions of the subcrust, and this picture was generalized to the whole crust by measurements of the earth's thermal gradient in mine shafts and tunnels. A greater density for the earth's interior was inferred from the discrepancy between the density of superficial, crustal rocks and the density of the whole earth, which was large.[1]

But nothing exact was known about the gradient of temperature or the distribution of densities within the earth, nor was the state of matter known. Even at the end of the nineteenth century the earth's interior was the true terra incognita; the behavior of rocks under temperature and pressure was not directly known beyond shallow depth. Geologists had tried various experimental simulations of conditions within the earth since the eighteenth century in order to demonstrate how temperature and pressure might produce various metamorphic and deformational effects, but most of these efforts had been directed to the replication of deformation and physicochemical alteration of sedimentary series and not to the

1. In fact, from the beginning of the nineteenth century the mean density of the earth was known to within 1 percent of its current value, a close estimate of the earth's ellipticity was available, and there was some theory on the propagation of disturbances in a deformable medium. See Bullen 1974:11.

behavior of the material of the subcrust.[2] Moreover, the variety of materials and the amounts of heat and pressure applied over different amounts of time were so varied as to render the enterprise inconclusive. From the time of Hutton till the twentieth century such experimental results were generally interpreted as suggestions of possibilities, and nothing more.

The picture of a collapsing crust around a shrinking interior and more particularly the idea of a strong and rigid crust that only occasionally, and then rapidly, gave way to the stresses accumulated during the retreat of the supporting interior mass, was not so much a deduction from physical quantities as a possible explanation of intermittent orogenic and marine cycles and of the kind of compression and faulting seen in ranges like the Alps. Although some attempts had been made to approximate more closely the process, as in Heim's work (1878) on the deformation of Alpine strata and the associated crystalline masses, the mathematical aspect of the explanations amounted only to estimates of the total crustal compression and total volume of erosion and to speculations on the relatively shallow depth at which rocks would be deprived of strength. Heim's work was directed only toward the establishment of the possibility and plausibility of the lateral translation of sedimentary masses by plastic deformation in response to the tangential expression of the forces of contraction. While his sections through the Alps mapped distances of 100 kilometers, he depicted only a minuscule fraction of the earth's radius in depth. Like the nappe theorists, he became in his later work increasingly divorced from physical considerations and devoted to tectonic analyses alone.

The general weakness of the nappe theory and the syntheses of Suess and Haug was their superficiality; they had little or no direct foundation in physics, being deductions from surface phenomena, visible structure, and the sedimentary and paleontological history of the crust. Each implied a theory of the whole earth, but only in the most general terms was the behavior of the interior specified. Lord Kelvin's support of the theory of secular contraction was welcomed by its exponents, but Kelvin was concerned above all with the age of the earth and the time that had elapsed since it had begun to cool. His deductions, based on only a cosmological hypothesis, the most general thermodynamic considerations, and loose and undefined assumptions about the behavior of rocks under great

2. No history of experimental geology exists, but see E. B. Mathews, "Progress in Structural Geology," in *Fifty Years Progress in Geology: 1876–1926* (1927), especially pp. 147–153.

temperature and pressure seemed to suggest that secular cooling might indeed be responsible for the corrugation of the crust into networks of folds.[3] The debate over the material state of rocks given certain distributions of heat and pressure was the source of the most contention among geophysicists and between geophysicists and geologists, and it provided the greatest challenge to the synthesis of geology based on the contraction theory.

Stated in the most general terms, the notion opposing the theory of contraction of the earth (insofar as that implied the progressive subsidence of the crust) was that the crust of the earth is in dynamic equilibrium with a substratum of mobile matter. This matter, though it might be slowly cooling, would still be able to respond to subsidence in one portion of the crust with an uplift of the crust somewhere else, as the mobile matter beneath was rearranged. This idea was the foundation of the hypotheses of Herschel in the 1830s. He had argued that an increase of sedimentary load on a portion of the crust must lead to local subsidence and heating and to an adjacent local uplift and cooling as the matter of the subcrust flowed away beneath. The notion provided both the "great desideratum of the Huttonian theory"—a source of heat to deform the strata—and a mechanism by which continental uplift might occur; sedimentary transport disturbed the equilibria of temperature and pressure in the crust and subcrust. The idea was not worked out in mathematical detail and was increasingly eclipsed by the success of the contraction theory, though some American geologists, like Hall and Hunt, remained partisans of the notion in conjunction with the Lyellian picture of uniform continental uplift.

The relations of the crust and the subcrust finally emerged in 1854 as a matter of concern in India. In 1847 the Great Indian Survey had published the results of the measurement of a meridional arc across the Indian subcontinent. Included in the results was the remark that a discrepancy had appeared between the astronomical and geodetic (triangulation) measurements of the distance between the cities of Kalianpur and Kaliana, 375 miles apart. As head of the survey, Colonel Everest was willing to distribute the error (which amounted to 5.236 seconds of arc) among the geodetic stations and dismiss it as an error of measurement (Markham 1878: 56, 140).

The matter might have ended there but for the presence of John Henry Pratt (1809–1871) the archdeacon of Calcutta. Pratt

3. For a full treatment of this topic, with the chronological implications, see Burchfield 1975.

was a Cambridge-trained mathematician (graduating in 1833) whose earliest scientific work had concerned the determination of the figure of the earth to an accuracy that would demonstrate the exact oblateness of the spheroid (Manheim 1975:126).[4]

In 1854 a long communication from Pratt arrived at the Royal Society, in which he expressed his concern for the resolution of the discrepancy noted by Everest.

> The importance of accounting satisfactorily for the difference between the geodetic and astronomical results appears from the effect it must have on the determination of the earth's ellipticity: an effect such, that unless the quantity can be fully accounted for, it must render the great Indian Survey comparatively useless in the delicate problem of the figure of the earth, however valuable it may be for the purposes of mapping the vast continent of Hindostan. [Pratt 1855:55]

Pratt's contention was that it was far more likely that the astronomical instruments were at fault than the geodetic measurements, and he argued that the discrepancy was probably due to the gravitational attraction of the plumb bob (used to level the instruments) by the mass of the Himalaya Mountains in the north; this mass had drawn the bob out of the true vertical and caused the resultant error. While, he wrote, the effect must also have been felt by the plumb bob used to level the geodetic instruments, the deflections would there be insignificant because the instruments were aimed horizontally rather than vertically (as the astronomical instruments were). The slight elevations and depressions from the horizontal (line of sight of the theodolite) between the forty-seven stations of the geodetic triangulation would result in a total error "not higher than the second order," whereas "in the expression for the inclination of the two verticals at the extremities of the arc they occur in terms of the first order" (Pratt 1855:55).

Pratt wished to discover if his argument (the attraction of the bobs by the Himalayas) could account for the direction and the amount of deflection. To do so, he undertook to calculate the mass of the Himalayas. "To dissect and actually calculate the attraction of the masses of which the Himalayas and the regions beyond are composed, appears, at the very thought of it to be an Herculean

4. His larger works, now rare and perhaps no longer extant, are J. H. Pratt, *The Mathematical Principles of Mechanical Philosophy* (1842), revised and expanded as *A Treatise on Attraction, La Place's Functions and the Figure of the Earth* (1860). See Manheim 1975.

undertaking next to impossible. I am fully convinced that no other method will succeed" (Pratt 1855:57). Pratt took the figure exercising attraction to be an irregular polygon bounding the mountains and plateaus of its various triangular sections at their average height above sea level. Using Nevil Maskelyne's estimate of the density of the Scottish mountain Schiehallion (2.75) as the density of the crust, Pratt calculated the total mass and arrived at an astonishing result: the observed error ought to have been three times as large, given the mass of the mountains. His laborious and lengthy calculations of the mass and volume of the northern end of the Indian subcontinent indicated that the deflection of the plumb bobs should have resulted in an error of 15.885 seconds of arc between the northern and southernmost astronomical stations.[5]

Pratt tried several stratagems to reduce this discrepancy. He lowered the estimate of density to 2.25, but the calculated deflection remained larger than the observed. He then additionally reduced the size of the total polygon of the Himalayas and Tibet by eliminating the "doubtful regions" known only from sketchy reports of travelers. He then reduced the median elevation of the known area by several thousand feet. The discrepancy remained in spite of the lower estimates of area, density, and altitude. Pratt was convinced that his calculations of the attraction of the Himalayas were "not far from the truth" but found no way to reconcile them with Everest's observed error (Pratt 1855:100). Perhaps, he suggested, there was something about India that made Everest's coefficient of the ellipticity of the earth too large for the Indian arc. He hoped that someone might further study the problem.[6]

Pratt's problem engaged the interest of George Biddell Airy (1801–1892) the Astronomer Royal of Great Britain. In a short paper to the Royal Society he offered a comment on the problem of the anomalous deflection. He had been at first surprised that the observed discrepancy was not as large as that calculated from the theory of gravitation, but he became "soon convinced" that the result should have been anticipated (Airy 1855:101).

Airy considered the following situation. He imagined the earth

5. Information on Maskelyne's 1774 estimation of the density of the crust and its relation to Bouguer's anomalous results in Peru at Mt. Chimborazo is contained in Bullen 1974:13–14.

6. While this paper is often cited as the origin of "Pratt isostasy," in which all elevations are seen to be deficient in mass, this idea nowhere appears in the paper. Pratt lowered the estimate of the mass of the mountains but also lowered their height and area and ended by seeing the discrepancy not as a measurement error but as an erroneous assumption about the ellipticity of the earth.

to be a spheroid with a crust 10 miles thick over an interior of "a fluid of greater density than the crust" (Airy 1855:102). He then supposed that upon this crust a tableland should exist 2 miles high and 100 miles broad "in its smaller horizontal dimension" and that this should be the only relief in the crust. "Now I say," he wrote, "that this state of things is impossible; the weight of the tableland would break the crust through its whole depth from the top of the tableland to the surface of the lava, and the whole or only the middle part would sink into the lava." To avoid such sinking, the tableland rock would have to have "cohesion . . . such as would support a hanging column of rock twenty miles long. I need not say that there is no such thing in nature. . . . If instead of supposing the crust ten miles thick, we had supposed it to be 100 miles thick, the necessary value [of cohesion] would have been reduced to $\frac{1}{5}$ of a mile nearly. This small value would have been as fatal to the supposition as the other." Every rock, he admonished, had mechanical clefts that rendered the adhesion of even the firmest rock, when considered in regard to large masses, absolutely insignificant (Airy 1855:102).

Airy's simple calculations opposed the notion that mountain systems could be merely great lumps of rock sitting on the crust, and he proposed instead the theory of mountains that has ever since been associated with his name—the so-called roots of mountains hypothesis, also called Airy isostasy.

> I conceive that there can be no other support than that arising from the downward projection of the earth's light crust into the dense lava. . . . It appears to me that the state of the earth's crust lying upon the lava may be compared with perfect correctness to the state of a raft of timber floating upon water; in which, if we remark one log whose upper surface floats much higher than the upper surfaces of others, we are again certain that its lower surface lies deeper in the water than the lower surface of the others. [Airy 1855:103]

Airy's argument was that the discrepancy between the calculated and the observed error was attributable to the fact that beneath the Himalayas and Tibet and any similar elevation, there was a mass, or root, of light crust extending downward and displacing the dense lava of the interior, which, though it could deflect the plumb bob, could not do so to the value expected if it were assumed that the Himalayas were an extra lump of matter on top of a homogeneously dense crust. Airy believed that the "deficiency of matter" below a major mountain system produced a negative effect, less attraction than would be supposed.[7]

Pratt's response to Airy's proposal came in 1859 and is important as the first statement of what came to be called the hypothesis of uniform depth of compensation, or Pratt isostasy (Pratt 1859). The theory was used as a model in the extremely important gravimetric survey of the United States conducted by John Hayford in the first decade of this century.

Pratt assumed with the Cambridge mathematician William Hopkins (1793–1866), his former teacher, that the crust was between 800 and 1,000 miles thick and that it was *more* dense than the substratum; both assumptions were contrary to Airy's. Believing the crust and the substratum to be composed of the same material and that cooling and contraction should lead to an increase of density, he thought the still fluid matter must be less dense than the solid crust (Lyustikh 1960:30). A crust of such thickness would have more than enough strength to support a mass like the Himalayas. Moreover, Pratt had a theoretical objection to Airy's hypothesis. It would lead, said Pratt, "to a law of varying thickness [of the crust] which no process of cooling could have produced." In Airy's model of the earth "every protuberance outside would be accompanied by a protuberance inside [which] would mean that there should be a depression under each hollow in the crust, including the ocean bottoms" (Pratt 1859:747).

Pratt subscribed to the doctrine of secular contraction in a thick crust. The crust was most depressed in the coolest portion; the mountains and tablelands represented the relatively warmer, less contracted uplands. Given the homogeneity of the crust, differing elevations evidenced different distributions of temperature, and consequently of density, down to a "level of uniform density" somewhere below. Imagining the crust as a series of columns of rocks side by side over the whole surface, he hypothesized that differences in elevation were achieved by different distributions of density down to a level of about 100 miles, where all densities were uniform and the differences in altitude "compensated."[8]

7. It is amusing to note that while Airy's name, because of this hypothesis, is invoked more by geologists than astronomers, mention of this paper occurs neither in his autobiography nor his biography in the *Dictionary of Scientific Biography*. The catalogue of his published papers reveals that he never returned to the subject or thought any more about it. For a man obsessed with his own fame—his journals, which record his thoughts almost hour by hour, run to more than 500 volumes—it is interesting that he seems to have measured this contribution lightly.

8. Actually, he computed the effects of various distributions of density between 100 and 1,000 miles of the surface on the attraction of plumb bobs and noted that

The question remained whether such a conjecture had any foundation in nature: to this, Pratt had no answer. Eventually, he decided that the most likely cause of the discrepancy was not the Himalayas at all but higher-than-normal density beneath the surface near some of the stations (Pratt 1859:762–763).[9]

The dispute, which was, after all, a way to correct technical errors in geodetic calculations, lay dormant for some years. The terms of it were revived in the 1870s by a British geophysicist, the Reverend Osmond Fisher, who was also interested in the figure of the earth and certain problems he had detected in the geological interpretation of the theory of earth contraction.

In 1876, Fisher pointed out that a number of geological effects associated with the theory of secular contraction were unlikely or gratuitous. He agreed that secular cooling led to the wrinkling of the crust in response to a diminished nucleus but argued that the deepening of the oceans was not a necessary consequence (Fisher 1876:325). Moreover, the great pressure of the interior dictated that the earth should cool as a solid; the notion of the foundering of blocks of crust into the interior was eliminated by the absence of any liquid or fluid medium into which they might founder (Fisher 1879:417). In addition, the reduction in volume of rock passing from fluid to solid state, which he calculated at about 6 percent, made it unnecessary for the crust to collapse or break up and sink as a gesture of accommodation to the diminished volume of the interior mass. Although the presence of mountains dictated a former fluidity of the subcrust, the time might have already passed when such process could act—an idea similar to Dana's notion that the crust would become progressively more resistant to the downward pull of contraction (Fisher 1876:330).

Fisher pointed out that all the phenomena of contraction were dependent on assumptions about the compressibility of rock. He hypothesized an imaginary "datum level" that would represent the smooth surface of a perfectly compressible crust that responded to pressure only by an isotropic increase in density and a smooth decrease in volume. That such was not the case he knew full well, but he argued cogently that all contraction expressed as folding

---

with a depth of uniform compensation as shallow as 100 miles, the error at Kaliana (the northernmost station of the arc) was reduced to 1.538 seconds. (Pratt 1859:759).

9. This suggestion was adopted by the Indian survey, which, after 1860, used pendulum observations of gravity at the astronomical stations to check "the tendency of the plumb-line to deviate from its normal direction, in consequence of local irregularities of the earth's crust. . . . " See Markham 1878:140. Pratt seems to have been as little aware of the source of his ultimate fame as Airy.

must occur as elevation *above* such a datum level (Fisher 1879:416). As the earth cooled and solidified any increase in the amount of folding would have to be expressed as an addition of matter, *above* the hypothetical level of perfect compressibility; the crust must, through time, become thicker and stronger. He revived the argument that mountain ranges need not be anticlinal folds but might be carved out of any elevated tract, because the effects of contraction ought to be uniform over the surface of the sphere.

He also essayed the idea that continental oscillation accorded with the idea of a contracting globe. Continents would be elevated when contraction was "relieved" under the oceans, whose bottoms would be uplifted when, as the additions to the continental bases solidified and cooled, the lateral movement of the "viscous couch" of the substratum pushed up the oceanic basins in turn (Fisher 1879:454).

In 1881, Fisher published the results of these and other investigations in *The Physics of the Earth's Crust*, in which he tried to show that the hypothesis of secular contraction was insufficient to explain either the amount of compression of sedimentary sequences observed or the inequalities of relief that existed in the crust. Fisher argued that whether one accepted Thomson's hypothesis (Lord Kelvin's solid earth) or Airy's (flotation of major features in a denser substratum), the amount of shortening that the earth's radius would require, if it were the cause of all folding, was more than 700 miles. By assuming that the oceanic crust was more dense than the continents, the radial shortening could be reduced by an order of magnitude, to a figure as small as 42 miles; this, however, was still too vast (Fisher 1883:76).

Fisher believed the Airy hypothesis to be true. He believed that the continental crust was made of acid (granitic, lighter) magma and that the oceanic crust was made of a more basic (basaltic, heavier) magma, but he could not explain how the crust could maintain an equilibrium of figure with such persistent density differences in the earth shell. Indeed, the entire picture of the distribution of continents and oceans was something he felt he could not explain: while contraction was "totally inadequate," an adequate hypothesis had still to be discovered.

When, in 1882, Fisher came upon George Darwin's theory of the creation of the moon by "fissipartition" from the earth, he saw it as an answer to his dilemma concerning density distributions. While Darwin had argued that contraction would soon heal and conceal the place at which the moon parted from the earth, Fisher argued that the ocean basins, and particularly the Pacific Ocean, constitute a scar of separation (Fisher 1882:243). The density difference between continental and oceanic crust existed because the matter torn off in the moon's partition was of the lighter continental variety, which, perhaps, had once covered the entire earth and now existed only as fragments of a former continuous crust. This accorded with Fisher's general picture of the earth as a series of concentric shells of increasing density. The outermost crust was solid as a consequence of cooling, the next shell was mobile under heat and pressure, and the nucleus was again solid, this time as a consequence of pressure.

Fisher's *Physics of the Earth's Crust* (1881) was read with tremendous enthusiasm by the American geologist Clarence E. Dutton, who gave it a most complimentary review in 1883 (Dutton 1883). Dutton had been working along similar lines, and he applauded both the general picture of the crust worked out by Fisher and the particular details: the insufficiency of contraction to explain folding, the general thickening of the crust, and especially the idea of hydrostatic equilibrium. Dutton warmly approved the picture of the flotation of the broader surface features of the earth—Fisher put the ratio of crust to subcrust at 0.905, about that of ice to water; thus the crust's flotation in areas of high relief was like that of an iceberg, with about nine-tenths of the total mass hidden beneath the "surface." Dutton noted that he had long been at work on the idea of such flotation (approximating the Airy principle) and had called it isostasy to express the condition of crustal flotation on a highly plastic or even liquid substratum.[10]

Fisher was most encouraged by Dutton's reception of his work and the suggestion that it had great use in geological theory. Defending his claim of the insufficiency of the contraction theory in *Nature* in 1883, Fisher invoked Dutton in support of his attack on the idea of secular contraction as the geological *vera causa*. Dutton had said that Fisher

> . . . has rendered a most effectual service in utterly destroying the hypothesis which attributes the deformation of the strata, and the earth's crust, to interior contraction by secular cooling. No

10. Every history of the concept of isostasy that I have consulted credits Dutton with the term and the idea in 1889. However, he published it here in 1882 with a claim that he had had the idea for several years; thus even in its modern form, the idea is at least ten years older than generally assumed. Since Fisher was at work on the idea of flotation of the crust as early as 1868, it is clear that the idea had been floating around since mid-century.

person, it seems to me, can sufficiently master the cardinal points of his analysis, without being convinced that this hypothesis is nothing but a delusion and a snare, and that the quicker it is thrown aside and abandoned, the better it will be for geological science. [Fisher 1883:76]

In 1886, Fisher responded to those critics who had pointed to numerous possible density distributions in the earth other than Airy's that satisfied both actual measurement and physical possibility. The principal alternative hypothesis was, of course, Pratt's idea of differently distributed density in a homogeneous medium down to a level of uniform compensation of relief as shallow as 100 miles. Pratt's version, retorted Fisher, was a fine and useful hypothesis for geodesy, but not for geology. The undeformed uplift of the Colorado Plateau notwithstanding, most elevated areas of the earth's crust were sites of great lateral pressure.[11] A theory of density distribution ought to be based on these areas, said Fisher, rather than on a single contrary example. He again defended the Airy hypothesis of roots, arguing that granitic rock, less fusible than basalt, could long survive immersion in a hot medium without melting away (Fisher 1886:10). He also defended the postulation of two distinct earth shells of different densities—a crust with a density of 2.68 and a subcrust with a density of 2.96—against Pratt's idea of differentially distributed density in a homogeneous medium down to 100 miles. A further weakness in Pratt's theory, Fisher implied, was that it excluded any deformation or development in the earth's crust. Pratt's inverse ratio of crust thickness to density could be maintained only where volume changes were the only changes—it would be fatally disturbed by lateral movements and other disturbances in the crust (Lyustikh 1960:31). Thus, Fisher's work moved closer to geology in the 1880s and away from purely geophysical calculation and interest.

A similar direction appears in the work of another Englishman, Thomas Mellard Reade, whose book *The Origin of Mountain Ranges* (1886) was, according to the author, the first treatment of the topic at length that had ever appeared in English (Reade 1886). Reade, too, was moved by the physical insufficiency of the contraction hypothesis. So many and so great, he wrote, were the difficulties of

11. John Wesley Powell's observations of the great undeformed uplift of the Colorado Plateau (published in 1875) "stimulated a strong preference for Pratt's explanation of isostatic compensation instead of Airy's. How could a region of undeformed strata bearing no sign of compression suddenly acquire a root of low density rock?" (Ursula Marvin 1973:50).

this hypothesis that several eminent geologists and physicists have assumed a molten layer between two solid ones. He referred explicitly to Osmond Fisher, who, he said, had followed out the theory of a collapsing outer shell to its logical end with a theory of mountain ranges supported by flotation in a denser medium. However, Reade opposed this theory and held for a solid earth down to the core (Reade 1886:7). He argued that whether there was a molten shell between the crust and the nucleus or the earth was solid throughout, mountain building by contraction of the earth was in either case impossible—the earth had not yet cooled sufficiently to achieve such effects (Reade 1886:131). Mountain ranges were the result of a completely different process and were independent of "vicissitudes of larger areal subsidences and elevations" (Reade 1886: 329).

The theory proposed by Reade for the origin of mountain ranges attacked both the physical basis of the contraction theory and the unification of geological thought based on the idea that mountain ranges directly resulted from long-continued subsidence in the crust contingent on contraction; thus he moved even farther from the contraction synthesis than had Fisher. Reade's theory, in spite of his claims of originality, is closely linked to the ideas of Babbage and Herschel, both dynamically and philosophically. Mountain ranges, Reade proposed, were the result of local temperature changes in the crust caused by very slow pulsations beneath. In an area of heavy sedimentation, a rise of isogeotherms in the crust led to expansion. Horizontal movement of the expanding matter being blocked by the earth's mass, the expanding material spent its force on itself—the sedimentary beds were folded, stretched, thrust, and elevated. Below a "depth of flow," rocks would release pressure by slow flowing in the solid state; this lateral "compressive extension" was also a cause of mountain ranges.

Reade argued that such phenomena occurred slowly and uniformly through small incremental changes resulting from small fluctuations in temperature over a long period: "ten changes of ten degrees would nearly equal in total effect a change of one hundred degrees (Reade 1886:329). Subsidences were seen as the result of a subsequent fall in temperature in the same region. However, each rise in temperature would leave a net uplift of ridged-up rocks removable only by erosion; and thus, mountain ranges.

Part of Reade's argument was his postulation of a "level of no strain" at relatively shallow depth, a peaceful hiatus between compressive strain above (the plications of strata trapped between colder masses) and expansive strain below (the zone of solid flow). Fisher, responding to Reade's attacks on his *Physics of the Earth's Crust* worked out a mathematical version of the theory of a level of no strain in 1888 and concluded that it was a strong argument against the older idea that cooling by conduction in a solid earth and the deformation of surface strata were intimately related; the existence of a level of no strain severed the connection of the crust from movements deep in the earth's interior (Fisher 1888:20).

Reade continued to publish and republish his theory, complaining that he was persistently misunderstood. He was particularly irritated by the assertion that he had revived the Babbage-Herschel theory, arguing that neither Babbage nor Herschel had ever given a theory of mountain building in any detail (Reade 1891:496). However, although a geophysicist like Fisher was attracted to the idea of a level of no strain, insofar as it constituted an attack on contraction theory, Reade's principal problem was that no one believed his theory of pulsations and temperature fluctuations or that the building of mountains by expansion was such a great advantage, as he himself believed. Bailey Willis, in 1892, stated the problem with Reade's theory most clearly.

> Contraction and expansion are not such widely different effects for any given change of temperature that the expansion of a zone a hundred miles wide could accomplish what contraction of a far greater arc shall fail to do. In that proportion by which the zone is shorter than the arc must the rise in temperature be greater to expand the zone by a given length than the fall in temperature required to contract arc by the same actual number of miles. Reade's argument seems to me to be either self-destructive or to suggest the cooling of the earth must have produced effects stupendous beyond all observed facts. [Willis 1893:275]

Reade's argument was conspicuously weakened by his admission that his theory of mountains could not account for the structure of the Appalachians; these mountains in the United States had the same role as the Alps in Europe: a theory that could not explain them had no chance of general acceptance.

Thus, in 1889, Clarence Dutton's attack on the contraction theory developed Fisher's line of argument and ignored Reade altogether. It is in the work of Dutton that the notion of dynamic equilibrium in the crust actually left the realm of mathematical geophysics and became a geological theory, the theory of isostasy, which was Dutton's name for the phenomenon. In an address to the Washington Academy of Sciences, Dutton dealt with the problem of the structure of the Appalachians and suggested possible theories that might explain their formation and character.

Dutton's principal argument was that no theory of earth shells of homogeneous composition, responding to contraction of the earth's interior by collapse of the outermost onto the receding inner shells, could explain the persistent trend of Appalachian folds along a determinate line, with overfolding always toward the interior of the continent. Moreover, the amount of folding observed was too great for any reasonable estimate of the possible contraction of the earth since it had begun to cool. "No rational inquirer can doubt that they have been puckered up by some vast force acting horizontally. . . . But the forces which would arise from a collapsing crust would act in every direction equally. . . . the hypothesis . . . is quantitatively insufficient and qualitatively inapplicable. It is an explanation which explains nothing which we want to explain" (Dutton 1889/1925:360).

Dutton noted, as had Fisher, that the earth must be inhomogeneous; only the regional accumulation of light and dense matter could cause the bulges and depressions of the surface; otherwise, the earth would be a smooth spheroid of rotation. Yet because the earth was a rotating spheroid, it must constantly seek an equilibrium of figure, irrespective of its homogeneity or inhomogeneity. "This condition of equilibrium of figure," he wrote, "to which gravitation tends to reduce a planetary body irrespective of whether it is homogeneous or not, I propose to name isostasy" (Dutton 1889/ 1925:361). The question of the approach of the earth's figure to a perfect isostatic equilibrium could *not* be solved by mathematical manipulation but only by recourse to geological evidence. While mathematical manipulation might establish limits of possible departure from isostasy, the actual condition of the surface was a matter to be explored by observation.

Arguing from the present appearance of the Appalachians and their sedimentary history—the tens of thousands of feet of shallow-water sediments deposited over great periods of time—Dutton proposed that isostasy was perpetually disturbed by erosion and sedimentation. The erosion along the continental margins caused a gradual subsidence of the crust along the line of greatest accumulation; simultaneously, the bordering upland plateau that was the source of sediments rose as it was disburdened of its load. Although the absolute elevation of surface features changed little, the continental borderland was able to supply erosion products to the subsiding littoral through great stretches of time.

Pendulum observations had revealed variations in the force of gravity that seemed to point to a greater density of the oceanic crust than the continents, since plumb-line deflections toward the sea were the most consistent observed deflections from true verticals. Dutton took this as positive evidence of an accumulated extra mass of denudation products at the continental margins (Dutton 1889/1925:360). This was crucial in determining the actual history of the Appalachians; he argued that under certain conditions, the disequilibrating transport of sediment to the littoral set up an underflow in the less (relative to the surface rocks) viscous substratum, which determined the direction of motion of isostatic recovery. The direction of least resistance in the substratum was inward toward the unloaded continent: "we may derive a force which tends to push the loaded sea-bottoms inward on the unloaded land horizontally." This force would explain not only the uplift constantly proceeding at the continental border but the origin of systematic plications folded toward the interior of the continent (Dutton 1889/1925:366). "The theory of isostasy thus briefly sketched out," he remarked, "is essentially the theory of Babbage and Herschel, propounded nearly a century ago" (Dutton 1889/1925:364–365).

And so it was. Dutton's argument, in the American context, was a continuation of the opposition of Hall and others to the theory of contraction put forward by Dana. He even reproduced Hall's argument that, but for erosion, some parts of the mountains would be 50,000 feet high. However, Dutton parted company with the older theory of continental uplift—the restoration of isostasy at continental borders was *not* responsible for greater regional uplifts, which in his theory, remained a mystery. The theory of isostasy offered no explanation for permanent changes of level: "On the contrary, the very idea of isostasy means the conservation of profiles against lowering by denudation (Dutton 1889/1925:368).

What isostasy did provide, for Dutton, was an explanation of mountain structure apart from contraction.

> Thus the general theory here proposed gives an explanation of the origin of plications. It gives a force acting in the direction required, in the manner required, at the times and places required, and one which has the required intensity and no more. The contraction theory gives us a force having neither direction, nor determinate mode of action, nor definite epoch of action. [Dutton 1889/1925:369]

In 1889, as the debate over the cooling of the earth intensified, Robert S. Woodward undertook a review of the newer mathematical theories of the earth and their combined assault on the theory of secular contraction. He traced the history of the contraction theory, noting that only recently had it been subjected to "mathematical analysis in anything like a rational basis" (Woodward 1889:350). He then looked at the theories of Reade, Charles Davison (who developed a theory of a level of no strain independently of Reade), George Darwin, Osmond Fisher, and Dutton and while noting their cogent objections argued that the theory of isostasy, or any other form of crustal equilibrium, was not, so far, as well set mathematically as the contraction theory, that its origin and adequacy was as yet undetermined, and that it would tend to an early and stable equilibrium of the crust unless it were disturbed by some exterior cause. He decided that all questions of contraction and crust mechanics—earthquakes, volcanism, the liquidity or solidity of the interior, the rigidity of the mass as a whole, and "whether the earth may not be at once highly plastic under the action of long continued forces and highly rigid under the action of periodic forces of short period"—awaited the resolution of our "profound ignorance of the properties of matter subject to the joint action of great pressure and great heat" (Woodward 1889:362).

Woodward's position nearly exhausts the logical (and historical) alternatives. One might accept contraction, accept isostasy, postulate some third process (Reade), or keep silent.

The only remaining alternative was eclecticism, and this was the approach taken by the American geologist Bailey Willis (1857–1949) in "Mechanics of Appalachian Structure" (1893). Willis approached the subject historically, observationally, and experimentally. His insistence on definition of the most basic concepts of faults and folding reveals both a wide acquaintance with contemporary European geology—the work of Heim and de Margerie, Lapworth, Geikie, Schardt, and others were represented—and a determination not to be misunderstood. His experimental work reveals a thorough knowledge of experiments with artificial sedimentary series from the time of James Hall to the most recent research of Alphonse Favre and Hans Schardt. He discussed the history of the investigation of the Appalachians and illustrated his work with full-color maps, structural diagrams, photographs, and tables of sedimentary series, taking particular care to establish the relative value of evidence previously collected by state geologists and the geologists of the Appalachian section of the United States Geological Survey (USGS).

At the end of the volume Willis put forward some theoretical suggestions. He observed, dryly, that the Appalachian province had been used to illustrate and "prove" a "majority of the conflicting theories of mountain origin" ever proposed and that "its broad facts have been appealed to by all who would generalize on the distribution and form of mountain ranges. . . . they have served impartially the dreamer and the student" (Willis 1893:274).

Although his observations and experiments showed that the Appalachians were the product of compression action from the Atlantic side of the system, the cause of the compression was a matter of open debate. Recently, doubts about the contraction theory had been raised by a number of investigators and theorists, particularly Reade, Fisher, and Dutton. While Willis found Reade's arguments of expansion as a cause of mountain building "self-destructive," the opinions of Dutton had been widely approved, and since they had been offered "the contractional theory has been less favorably considered than before the attack." Willis believed Dutton's argument to be "not well founded" and felt that it "must yield to reconsideration" (Willis 1893:277).

Willis knew, of course, of Dutton's dependence on the work of Osmond Fisher; Dutton had made no secret of it. However, Fisher, in Willis's mind, had "confused the lesser problem of zonal compression with the far greater one of the deformation of the spheroid" (Willis 1893:279). Fisher, and Dutton after him, had argued that contraction was quantitatively insufficient to explain folding. But, said Willis, considering the fact that both men admitted that they did not know the cause of the larger movements of the continents and the oceans, their exclusionary argument was of little weight: "Dutton's quantitative argument falls to the ground as at least not proved. It does not follow that contraction is quantitatively sufficient, but the question is still open." Dutton had also argued that the contraction theory was "qualitatively inapplicable." His qualitative objection had come in two parts: he had argued that the force resulting from contraction acted in all directions and not determinately, and he had said that the postulation that a great thickness of strata was a "zone of weakness" was a gratuitous hypothesis, designed to save the phenomena. The contraction theorists, he said, had not demonstrated why the force should act preferentially toward the continents nor why the thick Appalachian strata, and not the thin Atlantic floor, should be ridged up and plicated. Willis's answer was simple. If the force of contraction acts in all directions indiscriminately, then at least "the properly directed force cannot be denied its advocates" (Willis 1893:279). As to the argument that the sedimentary sequence was postulated ad hoc as a zone of weakness by the contraction theorists, Willis echoed the opinion of Thomas C. Chamberlin that it was not the thickness or thinness of a section of crust that determined its deformability but its attitude at the time a force was applied. If, as even Dutton had suggested, the strata were flexed down (or up), as on the slope of the subsiding continental margin, they would yield preferentially to more solidly established masses, said Willis, even if the latter were thinner. "When a bent strut yields at the bend, the locating condition is in the strut and not in the thrust" (Willis 1893:280).

The objections of Reade, Fisher, and Dutton were, in one sense, all of a piece:

> To every hypothesis brought forward to account for the folding of stratified rocks there is one objection made by its opponents: the cause is not quantitatively equal to the task required of it. For argument's sake, admitting for each and every one that the criticism is sound, I do not understand that it disposes of any [hypotheses] which are based on good inferences from observed facts. The problem of deformation [of the Appalachian system] is exceedingly complex and thus afforded opportunity for the action of more than one cause. As the work performed was stupendous, it required the combined power of all available forces. [Willis 1893:280]

Willis's argument was that the secular contraction of the earth and the isostatic readjustment of the littoral to a shift of load were complementary, superimposed processes. Contraction was an extremely slow and deep-seated process expressed over a great area; isostasy was more rapid and more local. Accepting the idea that isostatic readjustment might well, and probably had, initiated an underflow of the subcrust toward the continental margin, Willis argued that this underflow was accelerated by the secular contraction. During periods when the force of contraction was operative, the underflow was intensified and the coastal sediments, already at an unfavorable angle, were pressed inward against the resistant crystalline core of the continent and then severely folded and even overthrust. When contraction was "satisfied," the normal process of erosion and sedimentation, governed by the tendency toward isostatic equilibrium, began again.

Willis was eminently reasonable, allowing that any one theory was insufficient to explain all the phenomena and providing a sobering reminder that insistence on a single cause led generally to dogmatism rather than to a deeper understanding of phenomena. Yet, Willis's work was as much an attack on the synthesis emerging around the contraction theory in Europe as anything put forward by Reade, Fisher, or Dutton. It admitted that at some level, secular contraction was "quantitatively insufficient," but it left unanswered the central questions (which motivated many of the participants in the debate) of the ultimate mechanism, the ultimate criterion of structural correlation, and the reason for intermittent orogeny. Willis's notion that contraction was "satisfied" by a certain shift of mass and dislocation of strata suggested, without specifying, some idea of accumulating stresses in the crust, which idea was precisely the target of Fisher and Dutton, who argued that the process was extremely unlikely whatever the composition of the crust. Willis left these arguments unanswered, however, and his allowance for isostasy said nothing about the widely different conclusions one might draw about mobility if one accepted the Airy hypothesis or the Pratt hypothesis of compensation.

If anything, Willis's approach was that of a structuralist, a "tectonist," willing to accept the most *plausible* mechanism or combination of mechanisms as an explanation for observed structure but unwilling to have the phenomena of geology dictated ex cathedra by armchair geophysicists arguing for necessary responses of the crust to geophysical limits deduced from the results of geodesy, astronomy, and the physics of matter as currently construed. While formally unimpeachable and deliberately generous in his assessment of such results, Willis was ultimately unsympathetic to the concerns of either camp to know the actual state of affairs. He was satisfied with a reasonable and liberal approach to theory that provided the widest latitude for subsequent investigation, holding in reserve the vast fund of ignorance about the earth's interior as a clause against taking sides.

As this brief survey indicates, the number of hypotheses was rapidly increasing, and the protagonists were engaged in spirited debate over the relative merits of their versions. Willis's theory is an interesting variant on the theme of contraction and isostasy, but the very different thermal theory, proposed at the same time by Eduard Reyer (1892), was also an eclectic compromise. Reyer agreed that the contraction theory alone was not sufficient to explain folding but found geological problems with the major proposed alternatives. He admitted that the idea of a quasihydrostatic, or as he called it, "magmastatic," tendency of the crust to respond to loading and disburdening was a reasonable picture of a sufficiently plastic medium whose equilibria of pressure and temperature had been disturbed (Reyer 1888;1892). However, the earth was not "very plastic" and a number of geological facts could be adduced to demonstrate this: subsidence did not always proceed parallel to sedimentation; considerable sinking could be observed where the load was slight, and, "in many cases, enormous loading does not produce a depression of the earth's crust (volcanic chains growing upon a highland)" (Reyer 1892:224). Thus an automatic response to changes in load could not be assumed.

On the other hand, Reyer found the idea of a disturbance of thermal equilibrium as a "tectonic motor" eminently reasonable in most respects and approved the idea of a rise of geotherms leading to expansion and elevation in thick sedimentary sequences as physically "well-founded." It explained the tendency for mountain ranges to rise out of such sequences in shallow seas and not in the middle of continents, "which might as well occur according to the contraction-hypothesis." However, he saw no reason why, as in Reade's theory, the expansion should occur in a narrow zone: "there is no reason why the thermal expansion ought not to proceed through the rigid magma to the region of constant temperature" (Reyer 1892:224). That is, the expansion should proceed down to a depth of 500 or even 1,000 kilometers.

Here, there were geological problems. An unrestricted thermal expansion over a wide area would be a long, slow process, but this did not accord with the demonstration of relatively short and intense periods of mountain making. Further, folding had been shown to be a shallow process and to proceed in a horizontal direction, eliminating mere vertical expansion as the sole cause of mountain building.

To answer both objections Reyer suggested a unification of the thermal theory with his own theory of gravity sliding, which he had first proposed in *Theoretische Geologie* (1888). The slow uplift of a district of recent sedimentation would result in the lateral sliding and folding of large masses of sediment, even if the inclination were only between 5 and 10 degrees. That fold mountains often rested on an undisturbed base of crystalline rock was, he thought, a telling argument both against the contraction theory and against the idea of folding in response to thermal expansion in a restricted zone. An immediate objection to Reyer's theory, of course, was that "the hypothetical land [from which the folded sediments were pushed toward the lowland, the uplift] in the back of the chain is often wanting, and that in its place a (marine or a terrestrial) depression exists" (Reyer 1892:224). The presence of such subsident backlands figured prominently in the theories of Suess and the nappists and led both to a dependence on backland contraction as a way of driving the mountain systems of Europe to the north. Reyer's response was that a denuded backland would begin to sink as it cooled, and the continued evolution of a district would find a (thermal) backland subsidence as the natural successor to each epochal renewal of uplift, gravity-sliding, and cooling.

The hypothesis had some strong points. It explained unidirectional motion of folds (the direction of slide), the shallow depth of folding, the presence of backland subsidence, the repetition of folding in certain districts—all without recourse to the theory of secular contraction from which they had been deduced.

For a while, even Lugeon and Schardt had been open to Reyer's suggestions. At first, they were not concerned to demonstrate the mechanism of the motion of the Pre-Alps onto the foreland as much as to prove its existence. Later, with the demonstration of a "root zone," which corresponded to the squeezing up of great anticlinal folds at the rear of the Alpine system, and the seductive pressure for a synthesis of the entire Alpine system according to a determinate sequence of nappe thrusts driven forward by a great *traineau écraseur*, they rejected the idea of gravity sliding as a local response to uplift and adopted the more schematic, unified interpretation that is characteristic of the nappe theory.

Most of the geophysical theories of mountain building proposed in the 1880s and 1890s, whether thermal (Reade, Reyer) or isostatic (Fisher, Dutton), tried to separate the history of mountain building from the larger history of the globe by ignoring the relations of the mountains to the subcrust, by postulating a level of no strain at shallow depth, by seeing fold mountains as gravity-driven masses sliding off expanding uplands, or by postulating effects, like flotation and compensatory underflow, that were absolutely incompatible with a theory of mountains as a direct expression of general subsidence. While such theories did not uniformly deny the existence of a secular contraction from cooling, they did deny the tectonic significance the idea had been given by the theory of Suess. Thus, as a group, these theories challenged the synthesis of geology based on subsidence.

However, the theories opposed to contraction were also radi-

cally incompatible with each other, and those geologists who supported the tectonic efficacy of contraction wholly (Suess) or in part
(Willis) played them off against one another or simply ignored
them. All of the geophysical theories were notably weak in explaining the arrangement of continents and oceans. Dutton admitted
that the larger movements were a mystery. Fisher, while finding
the idea of subsidence as a cause of oceans "totally inadequate," was
driven to accept the partition of the moon from the earth as an
explanation for the existence of the oceans and for his postulated
density difference of continental and oceanic crust. Since the originator of that theory, George Darwin, accepted the idea of secular
contraction as an explanation of folding, and since Suess also accepted the idea of the fission of the moon, Fisher's claim could not
represent a fatal criticism of contraction theory. Reade had offered
only a theory of mountains and proclaimed the independence of
orogeny from larger movements, which he left unexplained. Reyer,
hostile to contraction as a cause of folding, argued only that uplift
and subsidence were the expression of a warping of the crust of
great amplitude and long duration and did not specify the reason
for the original arrangement of continents and oceans.

The merits of these various arguments against universally effi-
cacious contraction were recognized by the American geologist
Thomas C. Chamberlin, who tried to develop, within the bounds of
a general synthesis of geology anchored at the extremes of cosmology and structural detail, a unified theory of the earth that would
preserve the idea of periodic orogeny and marine cycles of transgression and regression. It was Chamberlin who tried to translate
the various contemporary and opposing streams into a new synthesis of geology, which would use the global frame of Suess's work as
necessary but would take into account the sound objections to the
older, simpler, and less reflective version of the theory of contraction as well as the newer ideas about crustal equilibrium suggested
by geophysicists and geodesists.

CHAPTER ELEVEN

Thomas C. Chamberlin and the Third Global Tectonics

In the 1880s and 1890s most American geologists had little time or inclination to undertake theoretical debates with their European colleagues. In the 1880s the first generation of surveys was just coming to an end, and for large sections of the country detailed maps had still to be prepared. The stratigraphical successions, particularly in the north central and western states and territories, still had to be worked out in detail, and the terminology was chaotic. More than 700 new stratigraphical unit names were introduced between 1886 and 1890, dividing the geological record into a total of 1,450 major known units. This did not mean that progress had doubled—duplications and missed correlations caused a third of these unit names to be abandoned later (Moore 1941: 183–196). While the number of names has now long since passed 10,000, the outpouring of new designations temporarily swamped any general paleogeographic attempts.

Thus the regularization of geological terminology begun at the first international congresses was of pressing interest to the American participants, faced with their mounting stratigraphical difficulties, and it was a source of profound irritation to them that so much time was given over by their European confreres to the dramatic declamation of important new discoveries and theories.

It is very natural that the eminent apostles of research who attend these congresses should wish to bring before their fellows the latest results at which they have arrived and that the great body of members should wish to hear them, but then excursions to the boundaries of the picket lines of science should first be relegated to extra hours, and secondly they should be few enough to enable all possible business which cannot be settled through the media of scientific publications, to occupy every precious minute of the time when the members are together, for the reunions are costly and difficult. [Frazer 1888:10]

Moreover, the Americans had their own debates. Between 1842 and the 1880s the "Taconic question" had been hotly and acrimoniously debated in the attempt to establish the timing of the first post-Archaean orogeny in the United States (Rice 1915:29). The question of rock metamorphism was another source of consuming interest, and further, James Hall and T. Sterry Hunt continued to press their version of the history of the Appalachian region in the early 1880s.

Practical and organizational considerations separated Americans from the debates of European geologists, but another reason kept them relatively insular, too—they had a general theory of their own. Dana's analysis of the evolution of North America was everywhere influential in the latter part of the nineteenth century in American geology, and his version of the contraction theory, stressing borderland growth of the North American continent in geosynclinal depressions and successive orogenies and annexations to the permanent core, was an adequate basis for American workers, particularly in the eastern United States.

Yet the destructive criticism of the contraction theory, advanced in the 1880s and 1890s by such geologists as Dutton, Reade, and Reyer, who favored isostatic or thermal theories of mountain formation, severely compromised Dana's integration of all the phenomena of geology in a single dynamic framework that explained mountain building as an aspect of the deep-seated movements governing the origin of continents and oceans. Opposition to contraction theory changed the content of tectonics and, further, challenged its central place in geological theory at the very time that European geologists were forging a synthesis of geology in which the primacy of tectonics was axiomatic. The most forceful critics of contraction were content to abandon geotectonics in favor of regional structural geology and implicitly adopted the position that theories of the highest generality were not an essential part of a sound research strategy.

The conviction that geotectonics ought to provide the framework for geological theory was never as strong in the Anglo-

American community as in Europe; and many North American
geologists were comfortable with James Hall's contention that geology was a simple and harmonious study that could inductively generate empirical laws in the absence of grand theory. Dana's theory
was acceptable and useful when its mechanism was unchallenged;
he had made peace with the doctrine of continental oscillation in
the postulation of permanent, growing continents increasingly resistant to the downward pull of contraction, and he obtained the
requisite elevations of mountain masses with his model of geanticlinal upwarps of great lateral extent. But once the efficacy of contraction to achieve these effects was forcefully challenged, the utility of the theory vanished. One might then adopt Dutton's isostatic
theory to account for the tectonic regime of the Appalachian province and accept the tremendous upwarp of the relatively undeformed Colorado plateau as a continental oscillation, dynamically
uninterpreted but graphically demonstrated. The major tectonic
provinces of North America thus accounted for, an attitude of
benign neglect in the matter of geotectonic theory seemed judicious and feasible.

This theoretical retreat, however circumspect, threatened the
autonomy of geology as a science by abandoning the ideal of a
complete planetary history of the earth. While this ideal was historically associated with the proponents of the contraction theory from
the time of Elie de Beaumont, it was not at all clear that a rejection
of a particular unifying dynamic entailed the rejection of all comprehensive theory. This point, suggested by Bailey Willis in his
criticism of Dutton, was extensively developed by Thomas C. Chamberlin (1843–1928), chairman of the Department of Geology at the
University of Chicago, founder of the *Journal of Geology*, and a
renowned glaciologist.

Chamberlin was convinced that the mere collection of observational data, which he called "the method of colorless observation,"
stifled research. Without a preliminary plan for synthesis under
some rational theory or some glimpse of the same, arduous fieldwork led nowhere (Moulton 1929:370). The "method of ruling
theory" was also destructive. A theory, once formulated, became
itself the center of attention, and investigators would use it to select
"important" data (which supported the theory) and to dismiss the
rest as insignificant. American investigators sinned in the direction
of colorless observation, and the Europeans did likewise in that of
ruling theory. Chamberlin's solution, and the most durable part of
his scientific legacy, was his advocacy of the method of "multiple
working hypotheses."

This method, as simple in concept as it was difficult in practice, obligated the field geologist to master the major candidate hypotheses concerning the problem under study and to carry this knowledge with him into the field. Properly practiced, the method should produce durable geological research; it would not allow local answers and ad hoc hypotheses that could not be integrated into any larger context, and it would not condemn a field study, and thus force the complete restudy of a given locale, if the ruling theory that it attempted to corroborate should fail. At the very least, this method would allow the reader to evaluate field studies in terms of his own theoretical commitments and to judge truly whether new results were absolutely incommensurate with whatever general theory he favored.

Chamberlin's large and widely used *Geology* was consistent with these views on the proper approach to geological theory. It was "comparative" in theory as well as in the usual sense of establishing generalizations by comparing structures and sedimentary sequences according to various standards: it recognized divergent hypotheses and carried out parallel schemes of earth history built on different principles. Frankly hypothetical, it directed the student's attention to the magnitude of unsolved problems (Chamberlin and Salisbury 1905–1907:I, vi).

While this was a sincere attempt, Chamberlin's own predispositions showed through. He had never advocated the stance of complete neutrality, a stance likely to produce the "colorless observation" that he opposed, and he confessed that in his text, which recognized alternative views, "the doctrine of continental permanence plays an unusually large part in the interpretation of continental evolution, of the migrations of life. . . . So also the doctrine of the periodicity of the great deformative movements forms a notable feature in the interpretation of life evolution" (Chamberlin and Salisbury 1905–1907:I, v).

This double commitment to continental permanence and to periodic deformation was an outcome of his own experience as a geologist, not of armchair theory. His role in establishing the geological history of the central United States, as chief geologist of the Wisconsin survey (1876–1882) and chief of the glacial division of the USGS (1881–1904), forced him to the early conviction that the North American continent had grown around a primitive core since Cambrian time. The periodic advance and retreat of continental glaciation had spurred his interest in the periodicity of all geological phenomena, including that of oceanic transgressions and of the relatively intense dislocations exemplified in the creation of mountain ranges. In 1898 he raised for the first time the question of whether these periods of dislocation, deduced from the stratigraphical record, were natural events or theoretical entities existing only in the imagination of geologists. "The most vital problem before the general geologist is the question of whether the earth's history is naturally divided into periodic phases of worldwide prevalence or whether it is but an aggregation of local events uncontrolled by overmastering agencies of universal dominance" (Chamberlin 1898a:450). The question had divided geologists in the 1830s in the original debates over the significance of the trends of mountain ranges. The point was not whether mountains were datable as discrete structures, but whether they could serve as chronological markers identifiable with a specific epoch in the history of the earth—and thus serve as a principle of correlation of formations in terms of structure and not lithology or paleontology per se. Chamberlin also thought that the evidence of marine cycles of transgression and regression needed to be considered as possible markers across continental boundaries.

Chamberlin's early (and persistent) conviction was that there was a natural, rhythmic periodicity in the life of the earth—"correlated pulsations," not "heterogeneous impulses" (Chamberlin 1898a:450). He believed that great earth movements led to worldwide readjustments of levels, that the oceans were deepening relative to the emerging continents, and that these phenomena were explicable by the periodic shrinking of the earth's radius, concentrated more in the ocean basins than on the continental masses (Schuchert 1929:330). He was equally convinced that the great epochs of mountain making were discrete and of relatively short duration compared with the great periods of baseleveling, in which the continental platforms were planed by erosion down to sea level —giving rise to marine transgressions. The earth alternated periods of long quiescence with periods of exceptional disturbance (Chamberlin 1898b:599).

These early convictions, which emerged as preferred hypotheses in the textbook of 1905, were promoted to the rank of postulates at the 1908 meeting of the American Association for the Advancement of Science (AAAS), where a series of connected essays on the paleogeography of North America was presented to the geological section of the association. These essays, accompanied by paleogeographic maps drawn by Bailey Willis, were published as a separate volume capped by a general essay by Chamberlin that encompassed the major outlines of his own theory: "Diastrophism as the Ultimate Basis of Correlation" (Chamberlin 1910).

Chamberlin began by noting the variety of views on the nature and cause of diastrophic (large-scale deforming) movements of the crust and by proposing to eliminate exact specification of mechanisms as long as general agreement could be reached. The question of whether movements were the result of contraction, expansion, or lateral pressure could be reserved as long as "we agree as to the general nature of their effects on the agencies at work on the surface of the lithosphere. We do not need to entertain the same conception of the nature of the earth's interior, if we are as one as to the working conditions which have prevailed on its surface" (Chamberlin 1910:298).

He then took it as given that the earth is divided into "abysmal" basins and continental platforms. Further, he dismissed dynamic interplay in the creation of deformation; on ocean bottoms forces were self-compensatory and did not affect relations of land and sea at the continental margins. Similarly, the deformation of continental interiors was also given as local.

Yet within this arena of agreement, there were still two possible positions. The first was that deformations, whenever they occurred, were strictly independent of previous deformations and could not affect cumulatively the size of continents and the depth and capacity of the oceans; further, they could not be made a trustworthy basis of correlation. The second was that major deformations were inheritances, following one another in "due dynamical kinship." The depression of the oceans and the elevation of the continents, while pursuing separate destinies, were nevertheless the expression of an overarching directional tendency of heavier (oceanic) and lighter (continental) masses of the crust. If this alternative were true, then we might expect renewed deformations, passing through the same phases in the same regions, as the continents *and* oceans tended to self-perpetuation. These two alternatives were the parting of the ways for the "interpretation of the larger events of geologic history" (Chamberlin 1910:300).

Chamberlin, of course, accepted the second view. The oceans and continents were permanent, through the diastrophic (large-scale deforming) renewal or "rejuvenation" of the continents. His argument was that elevated continents were worn down by river erosion to baselevel: streams cut away at the relief until their grade was exhausted and they would, or could, no longer flow into the sea. This exhaustion of relief marked the inauguration of a marine transgression and a facies change—that is, continental or terrigenous facies would be replaced by marine facies as the seas transgressed the baselevel of the continents; the signature of the reversal of the deposition cycle would be clearly evident in the reversal of facies, which, he asserted, was worldwide.

While this was a solution to the theory of marine cycles, it left the problem of the source of deformation to be considered. Chamberlin had tried to make his theory of baseleveling and transgression stand on the basis of stratigraphy alone, without reference to dynamical movements of any kind. To explain deformation and rejuvenation of continents and the restoration of differences in elevation between the continents and the oceans, he was finally forced to show his hand: the crust was strong and opposed deformation but would eventually build up stresses that would be resolved over a short period (of deformation) followed by a long quiescent stage in which new stresses would accumulate. The major means of the resolution of stress was subsidence of oceanic bottoms relative to the continents. While agreeing that some folding took place during deposition and accepting the possibility that erosion and deposition might set off local deformative movements, Chamberlin argued that these were not fundamental aspects of the larger movements. Seeing the earth as a solid spheroid and not a liquid globe with a thin sensitive crust, he postulated that the diastrophic cycles were actuated by deep-seated movements involving energies that rendered denudation and deposition trivial by comparison (Chamberlin 1910:306). He regarded the diastrophic rejuvenatory movements as ultimate in the sense that they were the means by which the stratigraphical and paleontologic record came to be—any general review of stratigraphic evidence on a continental scale "demonstrated" the reality of the alternation of baseleveling and diastrophic periods of vastly unequal length.

Whether one wishes to call this kind of theory "geological synthesis," "paleogeography," or simply "general geology," the subject matter is the same. Charles Schuchert, who recognized both Suess and Chamberlin as formative influences on his own views, stated, "What Suess did so well along descriptive lines, Chamberlin did even better along theoretic ones" (Schuchert 1929:329). While Suess's work exemplified the conclusions of a particular theory to a degree that makes a characterization of it as "descriptive" either misleading or erroneous, the comment places Chamberlin's general theory in its proper light.

Chamberlin had elaborated a new geotectonic hypothesis similar to that of Suess in scope but on a different physical basis, harmonizing a new version of contraction theory with an original isostatic hypothesis, the ensemble of which was supported by an entirely new cosmogonical hypothesis that Chamberlin had developed in collaboration with the astronomer Forest Ray Moulton.

This new cosmogony rejected the nebular hypothesis of Kant and Laplace that had been the foundation of the classical theory of earth contraction. The classical theory had been attacked for its inability to account for the amount of folding observed in the crust; its adequacy was further challenged by the contention that the nebular hypothesis on which it was based was itself physically flawed in that it could not account for the distribution of angular momentum in the solar system (Kay and Colbert 1964:673–674).

The new Chamberlin-Moulton cosmogony rejected the idea that the earth had formed in the coalescence of a spinning cloud of hot gas and proposed that the earth had been born in the "planetesimal infall" of cold solid matter and had grown by a process of "accretion." The initially cold earth, heated by the ever-increasing pressure on its interior as it enlarged, differentiated at some early period into relatively dense and relatively light fractions. These were not, however, concentric shells with the light material surrounding the dense, but a series of adjacent prisms. The oceans were dense prisms with their bases at the surface and their apexes deep within, and the continental prisms were reversed: the narrow apex of a continental prism protruded as a continental surface, while the wide base lay buried deep in the earth. This arrangement accorded with the distribution of oceanic and continental areas at the earth's surface.

Chamberlin's own version of the isostatic theory, which came to be called "Chamberlin compensation" vastly enlarged the areas over which compensation took place and applied it to the relations between continents and oceans. While the earth continued to contract under the influence of gravity, the progressive segregation of dense and light matter, itself a secular process, brought the continents and oceans steadily into isostatic balance.

Employing these concepts, Chamberlin worked toward a unified history of the earth. Whereas Suess had invoked a spasmodic negative eustatic movement, Chamberlin spoke of the rejuvenation of continental relief by gravitational contraction in the oceanic prisms and by massive isostatic readjustment. Dana's old idea of "arching pressure," the accumulation of stress at continental borders resulting in the lateral upwarp of continental areas of great extent, was preserved but with the difference that isostatic readjustments contributed to the pressure as well, the adjacent continental and oceanic prisms moving against one another; the borders of the prisms were treated not as mathematical fictions but real surfaces of compensation. The correlation of mountain formation and rapid oceanic regression based on deformation within or at the borders of continents followed in train (Chamberlin and Salisbury 1905–1907:II, 1655). Admitting that deformation and the creation of relief at continental borders resulted from a shift of load or at least was allied with the cycle of erosion and deposition, Chamberlin interpreted the process not as discrete (Dutton's borderland isostasy) but as an epiphenomenon of larger adjustments of continental and oceanic prisms.

The history of life was traced through the expansion and contraction of faunal assemblages. Faunas expanded during marine ingression over a wide, baseleveled area and contracted with the rejuvenation of continental relief, which progressively partitioned environments (Chamberlin 1910:304).

Chamberlin's theory departed in important particulars from the elements of competing candidate hypotheses. He did not discuss geosynclines in his diastrophic theory, and he treated orogeny with slight reference to lateral movements in the crust, concentrating on motion in the vertical plane. He denied continental foundering because, given his solid earth, there was no fluid interior into which continental fragments might founder. His cosmogony and isostatic theory made him more mathematical and explicit in the matter of mechanisms than Suess, Haug, or Dana.

While departing in particulars, Chamberlin preserved the scope of these alternative hypotheses, and the geological theory of diastrophic cycles emerged as a major geotectonic scheme, rescuing the major generalizations of the older contraction theory from the demise of its mechanism. Like Suess, Haug, and Dana, he defended a theory in which the oceans deepened through time; he supported the correlation of orogeny with periods of oceanic regression; and he agreed that there had been three major cycles in world history, even if smaller oscillations and episodal disturbances had left their marks in the interim. He insisted that these major cycles were not a chance collocation of regional events but a secular planetary tendency toward activity and quietude in different periods. This tendency was demonstrably global in scope and resulted from real, major rearrangements of the whole crust of the earth, even if the deformations were expressed again and again in the same regions.

One can see, in Chamberlin's synthesis, a dedication to the preservation of a unified history of the globe based on some form of the contraction theory but also a sensitivity to the geophysical critics. He demoted orogeny from its pride of place as the most obvious expression of subsidence. He denied *universal* subsidence, holding a theory of subsiding oceans and permanent emerging continents characterized by different densities, which he explained in terms of the planetesimal theory and "Chamberlin compensation." To explain the larger movements of the crust, he proposed a theory of crustal warping similar to Dana's early theory and, at least in outline, to Reyer's. On the other hand, he opposed the idea of borderland compensation by loading of the crust, principally because it interfered with the all-important theory of baseleveling, which was to account for oceanic transgressions. If profiles were continually preserved against erosion by uplift, it was difficult to see how baselevels could be achieved.

It became increasingly clear that the relative adequacy of the various specific hypotheses of mountain building and the more general theories of the evolution of continents and oceans could only be determined by a demonstration of the interplay of the crust and interior. Whereas Dutton had counterposed isostasy and contraction as incompatible versions of this relationship, Chamberlin had rescued contraction by postulating a unique form of isostasy, and the theorists of thermal contraction and expansion had provided an account of tectonic phenomena that avoided the issue entirely. As long as isostasy remained a theoretical entity, rather than a demonstrated or measured condition, the method of multiple working hypotheses was a necessity rather than a virtue.

Throughout the 1890s and the first decade of the twentieth century, however, actual measurements of the earth's gravity field, designed explicitly to test various forms of the theory of isostatic compensation, were carried out by European and American investigators. The scientists involved—William Bowie (1872–1940), John Hayford (1868–1925), Friedrich Helmert (1843–1917), George Putnam, Grove Karl Gilbert (1843–1918), Hervé Faye (1814–1902), and O. Hecker—carried out extensive measurements using pendulums, torsion balances, barometers, and the older method of comparison of geodetic and astronomical results for the same arc, based on differential deflection of the leveling plumb bob.

The results they produced, principally between 1895 and 1908, seemed to confirm the idea that isostatic compensation existed broadly over the face of the earth. Gilbert (1895) and Putnam (1895), Hayford (1909), and Hecker (1903, 1908) all produced surveys (Gilbert, Putnam, and Hayford in North America, Hecker over the oceans) that seemed to show that the earth was, at present, in a state of near-perfect equilibrium. The oceans were deep because the crust beneath them was more dense than the continents; the continents were elevated because they were lighter than the oceanic rocks.

The difficulties in interpreting gravity data achieved by such surveys were immense, as the number of corrections—for latitude, deviation from sea level, attraction to topographic features—were large and involved laborious computations to eliminate all the possible sources of deviations from the normal gravity value. It was therefore taken as a strong confirmation of isostasy that although the major surveys employed substantially different modes of correction, they arrived at the same overall result of a general compensation over wide areas of the earth. This constituted a severe blow to the idea of the foundering of continents as a cause of oceanic basins and to the idea of orogeny as an intermittent collapse of accumulated crustal stresses over a shrinking interior. Conversely, it spoke strongly for the continents and oceans as permanent features of the crust.

The survey results did not, of course, cause a stampede of geologists, and especially not structural geologists, to the banner of isostasy, nor did they cause general abandonment of the notion of earth contraction. Theories of isostatic compensation depended on a number of highly tendentious deductions about the interior of the earth that not everyone was prepared to accept, and their significance to the study of geological structure was long debated. Isostasy was a bastard theory—"a geologist's name for an astronomer's deduction from a geodetic fact"—and it was jeered at by skeptics who had observed the various twists and turns it had taken in its development (Read 1941:256). B. K. Emerson defined it as "a sort of hydrostatic equilibrium with the water left out and the equilibrium somewhat doubtful" (Adams 1938/1954:397). Harold Jeffreys noted that after the coinage of the term *isostasy* by Dutton, "the history of further work on the subject was curious. Geologists seem on the whole to have accepted Airy's mechanism, some even with exaggerations. Geodesists accepted Pratt's" (Jeffreys 1970:214).[1]

  1. Jeffreys's point is important. Hayford's gravimetric survey of the United States employed Pratt's idea that the earth is always compensated at every period by different distributions of density down to a level of uniform compensation. Hayford's survey was the most detailed, widely known, and influential. However, in employing Pratt's method he followed (approximately) the method of Faye, who had argued that a major correction, long employed in the interpretation of such data, should be dropped. The Bouguer correction, which Faye (and Hayford) re-

Nevertheless, the isostatic theory was established by serious measurements carried out by able and reputable scientists, and a response was called for. Incredible though it seems, through all of this, Suess had been doggedly pursuing his trend-line synthesis of surface features begun in 1875. In 1904 he published the final volume of *Das Antlitz der Erde*, in which he broke a long silence and gave his interpretation of the internal structure of the earth and of the current researches into isostasy.

Suess opposed the classification of earth shells proposed in the United States and postulated three different "zones or envelopes as determining the structure of the earth, namely the barysphere of Nife (Ni-Fe), Sima (Si-Mg) and Sal (Si-Al) (Suess 1904–1909:IV, 534). Basing his evidence on the mathematical calculations of Emil Weichert and the seismological studies of R. D. Oldham, he postulated a nickel-iron core (Nife) with a radius of 5,000 kilometers, surrounded by a shell of intermediate density, composed of silica and magnesium (Sima) with a thickness of 1,500 kilometers, which was covered in turn by a thin, light layer of which silica and aluminum (Sal) were the principal constituents and which made up the rocky crust. The "stratosphere," the younger sedimentary envelope, was formed almost entirely at the expense of the Sal layer, which was chiefly formed of gneissic rocks. It was characteristic that he should limit his discussion to petrologic (truly geologic) inferences, and in the course of his discussion he developed a magmatic theory that modified some of his earlier views.

However, his views of isostasy are more pertinent to the matter at hand. Suess, as might be expected, opposed the theory and marshaled all the knowledge at his command to show how the

---

jected, assumed that since any mass of rock between the measuring station and sea level exerted a gravitational pull on the instrument, the "anomaly" at the station should be reduced by an amount equal to the attraction of the mass. It was Faye's argument (following Pratt) that since the mass between the station and sea level is already compensated (since the whole crust is compensated), no such adjustment should be made. Eliminating this term in the correction had the effect of dramatically reducing the size of gravity anomalies, which till then had been very great, although it did not reduce them enough to overthrow the demonstration that the oceanic crust was relatively heavier. What it did was to install a geophysical prejudice against the Airy version of isostasy (which employed no such uniform density distribution down to a hypothetical level of perfect compensation) and to provide an "evidence" that the theory of mountain roots (extensions of light crust into the substratum) was mistaken. The "evidence" was, of course, a matter of a choice of assumptions and not a legitimate conclusion, but its effects were to be significant within a few years. In the meantime, it gave geologists, whose opinion on the inadequacy of the Pratt version of isostasy was already well established, a further reason to doubt the theory of isostasy, or Hayford's demonstrations thereof.

results obtained were contradictory or inconclusive. He opposed all forms of compensation, local or continental, as inconsistent with geological observation and insufficiently confirmed. He also attacked the contention that contraction was insufficient to explain the phenomena and offered a complex argument of the distribution of subsidences in mountain systems, in their foredeeps, and in the oceanic basins.

Two years later (1911), in a letter to Charles Schuchert, who had dedicated his *Paleogeography of North America* jointly to Suess and Dana, Suess began: "First I must confess myself a heretic in all regarding isostasy. I have in my last volume given the facts which cause me to doubt anything like a deficit in gravity beneath the mountains. . . . I . . . doubt likewise whether any sinking can be caused by loading. All these loads seem trifling by comparison to the magnitude of the planet" (Suess 1911:101).

Suess's characterization of himself as a heretic reveals clearly the direction that theory had taken in these few years toward a general acceptance of isostasy and isostatic compensation in one or another form. In 1909 he had written that "it is with extreme distrust that the geologist regards all attempts to apply the exact methods of mathematics to the subject of his studies" (Suess 1904–1909:IV, 601). This opinion was already no longer generally current at its time of writing; *some* mathematical consideration was becoming increasingly essential for any general theory of the deformational history of the crust, and a number of geologists of the younger generation began to acknowledge the fact.

The shift of opinion visible in the work of Bailey Willis throughout this period is a kind of barometer. Willis had traveled to Europe in 1903 and had delivered a modified version of his *Mechanics of Appalachian Structure* at the International Congress in Vienna, where he met and made excursions with Maurice Lugeon (Willis 1912). From Europe, Willis left for China to investigate the geology of the Asian continent, which was known almost entirely through the work of Richthofen. While he retained a tremendous respect for Suess and his work ("the axial direction of folds, the *Leitlinien* of Suess constitute criteria for the analysis of continental structure which is scarcely second in importance to unconformities and deposits"), he opposed Suess's interpretation of the structure of Asia in terms of the contraction theory (Willis 1907b:392). The Asian continent was not overfolded onto a subsiding Pacific foreland, as Suess had postulated, it was instead *underthrust* by the lateral spread of the floor of the Pacific Ocean, an idea that Willis had developed in the "Mechanics of Appalachian Structure" (1893). However, whereas in 1892 he had seen the effect of oceanic spread as an expression of contraction, by 1908, he saw it as a response to a quasi-isostatic process of "creep" as the dense ocean bottom spread laterally in response to gravity.

In 1910, in a review of isostatic hypotheses that found him in favor of a general compensation over oceans and continents, Willis remarked that contraction was an inadequate mechanism and that the old idea of arching pressure of the oceanic floor on the continental borders (a consequence of contraction first postulated by Dana) could no longer be maintained (Willis 1910b:393). His apostasy was as notable in the question of geologic cycles and in the discussion of tectonic movements, like those in the Alps. At the 1909 meeting of the AAAS, Willis argued for a separation of the history of the continents and oceans, which were, he said, permanent entities. "It is a mistake," he claimed, "to reason from the history of one to the history of the other" (Willis 1910a:245). He meant that the history of deformation in continental borderlands could no longer be referred to subsidences in the oceanic depths, a reiteration of his rejection of deformation of geosynclines as a response to arching pressure. The structural relations of geosynclines were intracontinental; of oceans, extracontinental. He mused critically where the ocean waters would have been if there had actually been a subsided Gondwanaland, Suess's Indo-African paleocontinent.[2]

While Willis continued to accept the idea of "diastrophic cycles" put forward by Chamberlin, his characterization of them was not nearly as grandiose or confident. Refusing even to name the epochs (Caledonian, Armorican-Variscan, and Alpine), he referred to them as "n, n + 1, n + 2" and opposed Chamberlin's idea of a natural periodicity responding to some inner rhythm (Willis 1910a: 246). Each of the diastrophic cycles was understood by Willis as

---

2. In the concluding portion of his fourth volume of *The Face of the Earth*, Suess tried to answer Willis's objections to his version of Asian structure, which threatened the whole scheme of virgations and syntaxes directed by the overfolding of foreland subsidences on the Pacific rim. He also seems to have been aware of this particular objection and thus adopted the idea of "juvenile waters," put forward by Osmond Fisher and others in the 1880s—the oceanic waters in this hypothesis were increased through the expulsion of hydrogen from the earth's interior which combined with atmospheric oxygen. See Suess 1904–1909:IV, 584. Suess did not comment on the effect such a postulation might have on his eustatic theory. Much in the final volume of his work indicates that he realized the failure of major aspects of his synthesis (see IV, 628).

simply a collocation of many epicycles, shorter periods of emergence and submergence, limited to a single ocean basin and the adjacent continents. The cycles of different regions were parallel but not contemporaneous: "major cycles of worldwide conditions are constituted by the coincidence of regional conditions" (Willis 1910a: 247). Orogenic districts were sharply limited by mechanical conditions of local deformation, and the proper unit of correlation was not the mountain system but the oceanic province—Atlantic or Pacific. Diastrophic intervals were the basis of correlation not because they were natural worldwide phenomena, but because they were *short* and easy to recognize. Finally, he pointedly rejected the idea of a Tethys, an idea to which he was never very sympathetic, by accepting the notion of a Eurasian geosyncline from the Pyrenees to the Himalayas but no farther west; this was a tacit rejection of transatlantic continuity.

Willis's opinions on the structure of the Alps continued this repudiation of every major element of the older synthesis. While not questioning the existence or the extent of overthrust, he challenged the idea of a *pli-nappe*, an overthrust anticline, as "hypothetical," "conventional," and "incorrect." He argued that the Alps had no unidirectional motion, and that the Pre-Alpine overthrusts were the result of a complex movement of thrust masses from the south and the north. The "roots" of the Wildhorn-Wildstrubel nappe were, in fact, "the southern and farthest advanced remnant of an overthrust from the northwest." The famous Dent des Morcles and Diablerets nappes were not nappes at all but "a single complex structure of the Scottish Highland type." The whole concept of superimposed nappes moving from the south, he said, "should be replaced by explanation based on intersecting major thrusts, minor thrusts and folds" (Willis 1912:2–9).

The nappe concept remained dominant in European geology for many decades thereafter, but the point is still made—the old synthesis that once stood unchallenged was now the target of geologists of international reputation. The entire edifice was creaking and groaning under the weight of sound interpretations deduced from different assumptions about the tendency of crustal deformation.

Haug noted this new theoretical pluralism in place of the old consensus in 1911 in the final volume of his *Traité de Géologie*. By 1911 there were no fewer than five theories of mountain building; three theories based on one-sided compression, a single theory of bilateral compression, and a single theory of expansion. The one-sided theories included the theory of contractional subsidence (supported by Suess, Deluc, Prévost, and Dana); the theory of isostasy (Dutton), and the theory of superficial folds (Reyer). The two-sided theory of compression was Heim's old theory of crustal shortening and the squeezing out of nappes. The expansion theory was Reade's, which Haug, like everyone else, rejected. In Haug's mind, the only two serious candidate theories were contraction and isostasy, both of which explained the nature of folded zones, the direction of folds toward the continents, and the recurrent activity of the same geosynclinal depressions. Haug, as he had in 1900, once again took the position that some combination of the two was the actual state of affairs, and he recommended consideration of Bailey Willis's "Mechanics of Appalachian Structure" (Haug 1907–1911:I, 516–534).

Haug's equivocal response to isostasy reveals his hazy understanding of the challenge posed by the theory to the synthesis erected around the various forms of the contraction theory. Chamberlin's attempt to play down the importance of the mechanism of uplift and subsidence wherever possible and elsewhere to accept both contraction and isostasy was partially successful, but it involved unique definitions of both contraction and isostasy and thus did not address the conflict directly. Suess and many of his European followers remained adamant in rejecting isostasy as a serious challenge to the synthesis of geology based on secular contraction and refused to admit mathematical considerations into their work. This refusal notwithstanding, by 1911/1912 the elements of the older synthesis had been called into question once again—particularly oceanic subsidence and orogeny as the release of accumulated stress—both were open to debate. Orogenic theory was no longer necessarily linked with the evolution of continents and oceans or with the theory of marine cycles. The theory of cycles was itself somewhat suspect, as Willis's remarks indicate, and opinion veered back toward the Lyellian idea of the chance collocation of regional events and, as in the theory of Reade, toward a notion of infinitesimal increments of change.

When Suess died in 1914, the sincere honors paid to his work were sometimes mixed with a feeling that it was somehow dated, a part of an earlier era:

> The honest critic must frankly admit that, great as is this masterpiece of geological generalization, it suffers from two rather serious defects. Its author was almost too clever as advocate and

parliamentarian, and was, moreover, not without bias. With a
manner altogether masterful, he could dismiss as it were with a
wave of the hand important evidence which was unfavorable to
maintenance of his thesis, and, with equal ability, could magnify
the weight of much less valuable and important observations.
[Hobbs 1914:814]

The appearance of the final volume of *The Face of the Earth* had
been greeted with international acclaim: the 1910 meeting of the
Geological Society of America sent Suess a telegram of congratula-
tion signed individually by the entire membership of the association, and the Geological Society of London did likewise. But it was
recognized in other contexts as a work of theory as much as a
compendium: "Suess . . . had used the materials to achieve ends of
his own, bringing forth conclusions which the individual worker
did not perceive." And, "It may be noted, however, regarding the
mode of treatment, that the work is built upon an exhaustive study
of areal [sic] structural and paleontological geology. It sums up,
therefore, and uses with great power, the modes of research which
were especially employed in the *nineteenth century*" (Barrell 1910:
269–270).

Joseph Barrell's respectfully critical valediction is a measure of
the recognition of the need to accommodate sophisticated geophysical and mathematical reasoning in any theory of the earth. Suess
had labored to bring geology together and had succeeded in doing
so, but his very concept of the science, of its aim and extent, was, in
the early twentieth century, a dead issue. Lyell's theory of continental oscillation was overcome by Suess's enormously sophisticated
demonstration of the diversity of movements in the crust and the
importance of the history of dislocation and deformation as a
superimposed sequence upon the secular cycles of erosion and
deposition. That part of Suess's theory was valued and the results
neither challenged nor outmoded. But the mechanism of contraction, Suess's unifying thematic for the various processes, was demonstrably not adequate. As a general theory unifying all the phenomena of geology, it failed to account for too much of recent
work. While the theory would have its partisans in succeeding
decades, its collapse was as catastrophic as the phenomena it explicated.

Haug's fuzzy perception of the issues removed his grand synthesis also from consideration as a successor in Europe. Dutton's
isostasy was discredited, Dana's general theory outdated. Reade
failed to gain an audience, and he, like Reyer, openly admitted that

his theory was orogenic and not geotectonic. Chamberlin, using his immense influence, continued to promote his own general theory, but its cardinal precept of correlated pulsations failed to gain the assent of his most influential collaborator, Bailey Willis; the insistence on the secular permanence of oceans guaranteed that it could not be a successor theory in Europe. The correlation of paleontologic data with his diastrophic intervals, which Chamberlin thought "immeasurably strengthened" his interpretation of the continental borders, turned out to be one of the weakest parts of his theory. (Moore 1941).

The career of geology in the following decades confirmed the worst fears of the theorists who had striven to maintain coherence in geology through grand geotectonic syntheses. The rapid proliferation of subspecialties and their concomitant isolation from one another made impossible the kind of syntheses enhanced by wealth of detail and breadth of mastery that had been achieved by Suess, Chamberlin, and Haug. Even Lyell's confident picture of geology as a science in which the basic sciences would be brought together (which the geotectonic theorists had absorbed and employed) became a manifest impossibility: geology in 1910 was a pyramid with a rapidly expanding base but without an all-seeing eye at the top.

CHAPTER TWELVE

Radioactivity, Continental Drift, and the Fourth Global Tectonics, 1908–1912

By 1910 it was clear that the secular contraction of the earth could no longer serve as a general explanation of the structure of the crust or of the origin of continents and oceans. No general theory, however, was available to replace it. The basic concepts out of which a theory might be fashioned remained loosely and sometimes contradictorily defined. There were two distinct cosmogonical hypotheses—the planetesimal and the nebular. Attention to the cycle of sedimentation and erosion at the borders of the continents had resulted in two very different theories of the geosyncline—the theory of Dana and that of Emile Haug. The theory of isostasy, and more generally that of the equilibrium of the crust, existed in several different versions (Pratt, Airy, Dutton, Chamberlin) with different tectonic consequences, different modes of action, and different scope. All of these elements were combined and recombined, with varying logical consistency, by those geologists who had not abandoned the search for a comprehensive geotectonic theory.

Into this world of all-too-multiple working hypotheses there emerged a new and entirely unanticipated consideration—the discovery of radioactivity in the crust. In 1908 the Irish physicist John Joly (1857–1933) presented a paper, "Uranium and Geology," to the British Association for the Advancement of Science, and in 1909 he expanded it into a treatise, *Radioactivity and Geology* (Joly 1909). Joly's interest in the thermal properties of rocks and minerals and in the age of the earth long antedated the discovery of radioactivity (1896); it was natural that he should have been among the first to realize its potential impact on geological theory. He was, he said, well aware of the general conservatism of geological circles and the strangeness of radioactive considerations for many practicing scientists in different disciplines. He was also aware that many would charge that a synthesis of geology based on radioactive heating was premature; the science of radioactivity was in its infancy, and the geological community had already been severely buffeted by geophysical considerations that it was not well able to comprehend. Joly insisted nevertheless that the ubiquitous presence of a nearly eternal source of heat in the crust should be immediately explicated. The contraction theory, in particular, must succumb to the discovery of an independent source of heat that invalidated all calculations of secular cooling from an original hot state by simple conduction (Joly 1909:100).

With the discovery of radioactivity in the rocks of the crust, Joly found himself, rather suddenly, in possession of a general theory of crustal dynamics that might replace the contraction theory. This allowed him to strike a telling blow at the calculations of the age of the earth that had been proposed by Kelvin based on the theory of contraction; Joly's own estimates of the age of the earth, based on his calculation of the rate of increase of the salinity of the oceans, were far greater than Kelvin had allowed. The amount of heat produced by radioactive decay slowed the rate at which the earth might cool and even suggested that the earth might be heating up.

Applied to the theory of crustal structure, particularly that of mountain ranges, the idea of radioactivity provided an answer to how sediments in geosynclines might be deformed and uplifted. Reade's idea that thermal expansion would be expressed in uplift and deformation of thick sedimentary sequences had been attacked by both Reyer and Willis on different grounds—neither scientist could understand why the heat should be concentrated in geosynclinal depressions and not distributed over a wide region (Willis) or generally within the subcrust (Reyer).

Joly argued that the sedimentary cycle of erosion and deposition in geosynclines at the margins of continents functioned as a "convection of energy," leading to the accumulation of large amounts of radioactive material, which heated the geosynclinal sediments independent of any subsidence under load and rising isogeotherms. Dutton had urged that the notion of geosynclinal sediments constituting a "zone of weakness," more vulnerable than crystalline rocks to the action of tangential stress, was a gratuitous hypothesis invoked to "save the phenomena" for the contraction theory. Joly agreed that the notion of a zone of weakness, first proposed by Herschel, was wrong. If anything, the opposite was true: solvent denudation removed the very alkaline silicates, which conferred fusibility and made the sediments *more* resistant to deformation. However, the presence of radioactivity rescued the basic concept of orogeny as transverse compression of such sediments by providing the necessary additional heat to render the sediments pliable (Joly 1909:102–103). Given the local accumulation of radioactive material in geosynclines, the "geotherm of plasticity" would rise rapidly with increasing thickness of sedimentary accumulation, and the portion of the rigid crust bearing the increasing stresses would be consequently thinned; the result would be faulting and overthrusts at the surface, with the stress resolved below hydrostatically (Joly 1909:109–110).

Joly applied this theory to other geological problems, including the history of the continents. In his theory, the continents were permanent features of the crust and had been growing since the Paleozoic era by marginal accretion of infilling geosynclines, which were uplifted into mountains and progressively eroded seaward into the advancing troughs. He even applied the radioactive theory to a defense of the "new Alpine tectonics," the nappe theory, for which he thought he had found a convincing mechanism. Joly was aware of the history of the nappe theory and the history of Alpine investigations in general—the work of Studer, Suess, Heim, Bertrand, Lapworth, Lugeon, and Schardt. He was also aware that the theory had never found an adequate mechanism and that the idea of great recumbent anticlines sat ill with many structural geologists, since it was so "very different from generally preconceived ideas of mountain formation" (Joly 1909:143). He applied the theory of radiogenic heat to one of the more embarrassing problems of Alpine tectonics, the *déferlement* ("breaking") of the great sequence of northward-moving nappes, in which the synclinal flexures remained nearly motionless while the anticlinal limbs of the hypothetical folds were drawn out for great distances over the whole assemblage.

Joly rejected the idea that crystalline massifs, the horsts of Suess and others, could obstruct the movement of the nappe front and cause the overthrusting observed. The increase of temperature and decrease of viscosity with depth would extend to them as well as to the sedimentary sequences below the "depth of flow." He argued that radiogenic heat caused the upward migration of isogeotherms, thinning the elastic layer of the crust, and raising the "isogeotherm of plasticity." The result was that the hydrostatic zone extended into the synclinal limb of each nappe as it was overthrust. As a fluid, it would no longer accept translatory motion directly and would remain almost stationary relative to the advancing anticlinal limb above it. Moreover, the motion of the anticlinal limb would be facilitated by the hydrostatic pressure of the newly fluid rock below, and the uplift would aid the movement of the nappes.

Apart from the local problem of Alpine tectonics, the theory of accumulation of radioactive materials had other, broader consequences. Accepting the newly offered planetesimal hypothesis of an earth formed in the gravitational accretion of cold material from space, Joly argued that the inhomogeneous distribution of radioactive material would lead to local heating and a true, fluid convection; radioactive material would be transported to the surface of the globe, leading to a continual enrichment of the crust with radioactivity. In the long run, this theory established the inevitability of cooling and contraction of the interior of the earth on a time scale commensurate with the half-life of uranium but not with the adiabatic cooling from incandescence of a once-molten sphere, and therefore, cooling was no longer tectonically significant.

Joly's picture of the influence of radioactivity on geological science was governed throughout by a concern with the full range of contemporary problems—from cosmogony to orogeny; and he was preoccupied with the discovery of a generally adequate hypothesis that might attain the explanatory power and organizing force of the recently deposed theory of secular contraction. However, as a physicist and a supporter of the idea of continental permanence, he was concerned with mechanisms whereby the continents might grow "in place" and attain their structural character and elevation relative to the oceanic depths. He was not particularly concerned about explaining, or reinterpreting, many of the phenomena of intercontinental connection and correspondence that had been adduced as evidence for the geophysically discredited notion of sunken paleocontinents once thought to span the Atlantic and the Pacific oceans.

The notion of crustal radioactivity did seem to eliminate the older theory of secular contraction as a serious geotectonic hypothesis and to accord with (if not exactly support) the doctrine of continental permanence and, to a lesser extent, Chamberlin's planetesimal hypothesis. Continental permanence as espoused by Willis, Chamberlin, Joly, Dutton, Reyer, and other more "geophysically minded" geologists amounted mostly to the assertion that the continents had always been continents, that the oceans had always been oceans, and that there was no interchange of the two in geological history. The seas had transgressed the continents, to be sure, and retreated from them again and again, but the abysses had never been platforms above the surface of the water nor had the platforms of the continents ever been at abyssal depths.

Stated in this way, one can see that the idea emerging in the first decade of the twentieth century was not so much one of continental as *oceanic* permanence. Few since the time of Hutton or Deluc had proposed a full alternation of the two—Suess, Haug, and others had argued for a measure of continental growth. The difference between their theories, and the newer geology, was the older idea that the oceanic basins were created by the progressive subsidence of continental crust into the abyssal depths. In Suess's theory, this doctrine of subsidence was a fundamental principle explaining continental structure and marine cycles alike; in Haug's theory it was a late and epiphenomenal occurrence, required by his idea that the history of the earth was mostly the interaction of broad continental platforms and narrow geosynclinal seas. However, both theories opposed the notion that the oceanic and continental areas represented a fundamental and original differentiation of portions of the outer crust according to the density of their respective materials—a conclusion supported by the theory of isostasy, founded on gravimetric observations over both continental and oceanic areas.

As early as the 1880s there had emerged the idea, entirely apart from the theory of secular contraction and continental foundering, that although the continents were permanent, the oceans might not be. Osmond Fisher, following George Darwin, had suggested in his *Physics of the Earth's Crust* that the "fissipartition" of the moon had created the Pacific Ocean, and he had added that the continents might have drifted apart, sliding toward the Pacific depression created by the departure of the moon. He did not, however, provide a geological elaboration of the idea.

Such an elaboration was taken up by William H. Pickering in 1907. Pickering assumed that the moon had parted from the earth and desired to find out where, when, and the nature of the possible consequences that might have ensued. He acknowledged that "it is the general opinion among geologists that the continental forms have always existed—that they are indestructible." But he wanted to know the reason for their arrangement and existence—something upon which contemporary theory gave disconcertingly varied opinions (Pickering 1907:28).

Pickering noted what had long since been established by Suess and others: the Atlantic and Pacific coasts were different. The Pacific was girdled by long chains of mountains and volcanoes, generally convex toward the ocean; the Atlantic coasts were "low, flat, and composed of curves as often concave as convex." He also noted that the earth's surface exhibited two preferential levels; again, something known since Sir John Murray had produced a bathymetric chart of the oceans from the results of the *Challenger* expedition, which had been widely publicized in the 1890s by G. K. Gilbert and others. This distribution of levels, approximating the continental platforms and the oceanic abysses, had seemed to be an argument for the long-standing separation of the continents and oceans.

Pickering combined these facts with the notion that the earth had condensed from its "nebula" in concentric shells of different densities. When the moon finally tore away from the earth, "three quarters of this light crust was carried away, and it is suggested that the remainder was torn in two to form the eastern and western continents. These then floated on the liquid surface like two large ice-floes" (Pickering 1907:30). Pickering believed that all this had happened while the subcrust was molten and that the westward transit of the Americas took place on an ocean of fiery magma. The Pacific was the scar left by the moon's departure and the Atlantic was a rift produced as the Americas were dragged away by the moon's gravity. To support this notion, he produced a map of the parallelism of the Atlantic coasts.

Another general treatment of continental permanence and oceanic impermanence, this time not associated with any cosmic catastrophe, was proposed in 1910 by the American geologist Frank B. Taylor (1860–1938). In 1898 Taylor had argued that the earth had *captured* the moon in geological time, and the increase of rotation had driven the continents toward the equator; later he abandoned the idea (Aldrich 1976:269–271). What he continued to maintain was the central doctrine of Suess concerning crustal movements— especially the southward movement and compression of the Eurasian continent.

Taylor's theory, presented in an address to the Geological Society of America in 1908, was the first reasonably complete and modern proposal of the theory of continental drift as an explanation of the origin and distribution of the continents (Taylor 1910).

He hypothesized that a change in the oblateness of the spheroid of rotation had caused a dispersal of the land masses, once concentrated at the earth's poles. In the Northern Hemisphere the land had begun to "creep" toward the equator, and the same had happened in the Southern Hemisphere. However, the process, which began in the Northern Hemisphere, had shifted the earth's center of gravity to the south and weakened the dispersive forces in the Southern Hemisphere. This explained why the South Pole should be covered by a huge continent and the Southern Hemisphere be watery; conversely, it explained why there should be an ocean at the North Pole in a hemisphere covered by land masses.

Taylor's geological evidence was drawn entirely from a close reading of Suess's *The Face of the Earth,* and he saw his work as a modification of Suess's theory that would preserve Suess's plan of the surface of the crust and its tendency to deform in certain directions, as in the great southward thrusts that had created the Eurasian mountain girdle.

Taylor, like most geologists in this period, had seen the difficulties in the contraction hypothesis that Suess had upheld; but he opposed also the modifications of that hypothesis proposed by Bailey Willis and Thomas C. Chamberlin (Taylor 1910:180, 225). All forms of the contraction hypothesis, he said, had explained the deformation of the Tertiary mountain belt in terms of some force acting from the oceanic side, whether a general subsidence, a consequence of planetesimal readjustment, or some other deep-seated movement of the ocean floor. Why, he asked, should the cause of the deformations not lie on the continents themselves? This cause, of course, was the "creep" toward the equator of the continental crust fleeing the poles.

Taylor held fast to the principle of trend-line analysis in developing his picture of crustal movements. While praising the worth of transverse sections, he quoted Suess extensively to argue that they tended to exaggerate local details at the expense of the further development of the general plan of the trend lines, "the task which now awaits the geologist" (Taylor 1910:189). Yet to produce a new theory from Suess's synthesis, Taylor had to take some liberties with Suess's overall scheme. The northward movement of the Alpine system, which was the germ of Suess's entire conception of lateral movement of the crust, Taylor dismissed as an unfortunate departure from the successful picture Suess had developed in Asia. He argued that somehow Suess had lost sight of his own overall plan. The Alps, Taylor maintained, were backfolding against an underthrust from the north (an implicit agreement with the ideas of Bailey Willis).

In fact, wherever Taylor found overfolding to the north in the Northern Hemisphere, he reinterpreted it as backfolding against a south-directed underthrust. This was particularly required to include portions of western North America in the overall picture. Thus far his theory was a continuation of Suess's geological plan, supported by theoretical ideas about the changing shape of the earth, which Taylor claimed he found, in part, in the later volumes of Suess's own work.

It was in his treatment of the Atlantic that Taylor departed from Suess and from other current theories. The mid-Atlantic ridge struck Taylor as one of "the most remarkable and suggestive objects on the globe" (Taylor 1910:216). Taylor was astonished that it should maintain a medial position in the ocean for 9,000 miles. It was probably, he thought "a submerged mountain range of a different type and origin from any other on the earth." To explain it, he suggested that the ridge was a "sort of horst-ridge—a residual ridge along a line of parting or rifting—the earth-crust having moved away from it on both sides" (Taylor 1910:217). He postulated two separate episodes of drift. Africa had moved east before the Mesozoic, and remained an undeformed tableland in the Tertiary. South America had moved somewhat later, since the Cordillera of the west coast of the continent was an important part of the Tertiary mountain girdle. Taylor's explanation was vague however: "Their present forms and relations suggest that the force which parted them was one that tended originally to crowd the two parts toward each other. . . . The release of strain was found by a great diagonal fracture along which the crust divided in two parts that crept away in opposite directions. The mid-Atlantic ridge remained unmoved and marks the original place of that great fracture." Why Africa should creep away at the end of the Paleozoic and South America at the end of the Mesozoic was not further explained.

This entire description of drift, while an honest concession to phenomena that suggested the creation of the Atlantic in geological time by rifting, occupied only three pages of a fifty-page memoir devoted to the movement of continents away from the poles and the creation of oceans at various other sites by rifting—of Greenland from Canada, and on all sides of the Indian Ocean. Moreover, Taylor declined to speculate on ultimate causes and at the end of the study proclaimed that "for a change in the degree of oblate-

ness, either in oceanic oscillations, or in deformations of the lithosphere, one is inclined to reject all internal causes and to look for
some tidal force as the only possible agency" (Taylor 1910:226).
This meant rejection of isostasy, planetismal readjustment, or any
other subcrustal mechanism, and thus most of the work done in
geophysics in the preceding twenty years. His frame of analysis was
exclusively tectonic, even to a greater degree than that of Suess
himself. But like Suess, his work seems to be a monument to the
kind of analysis that brought forth the best efforts of the nineteenth century. The deafening silence that accompanied the publication of his theory is ample evidence of the concordance of his
colleagues with that view.

At one level, the works of Pickering and Taylor represent two
extremes of the possibilities of drift theory: Pickering's approach
was geophysical and his concern was with a period in the history of
the earth so distant as to be practically cosmogonical. Taylor, on
the other hand, was quite aggressively geological and his interest
above all was the most recent deformation of the crust—the great
Tertiary compression. In each case, westward drift of the continents was a subsidiary effect of some greater force (partition of the
moon) or simply an unknown process.

In 1912 another version of drift theory appeared, that of the
German geophysicist Alfred Wegener (1880–1930).[1] A number of
important points separate Wegener's work from that of Taylor and
Pickering, but perhaps the most important difference was his realization of the seriousness of the impasse created by the fall of the
contraction theory and the pressing need for a new working hypothesis in the earth sciences that would attend with equal seriousness to geological and geophysical evidence.

Wegener saw a widening rift in the Atlantic that had nothing to
do with continental drift—a rift between European geologists, still

1. A great deal of nonsense has been written concerning Wegener's actual
profession. It has been extensively argued that he was not a geophysicist but a
meteorologist and that this was in some way a reason for the rejection of his theory
when it came to be internationally debated in the 1920s. Wegener was already the
author of a textbook on the thermodynamics of the atmosphere at the time he
proposed his version of continental drift. His writings reveal him to be thoroughly
at home in the treatment of geophysical data, as comfortable, certainly, as any of
his contemporaries. He wrote throughout his life on geophysical subjects, and his
articles were published in geophysical journals. Eventually he held a chair of geophysics and meteorology at the University of Graz. His ideas were considered important enough to be discussed by geophysicists in every country in the world. If
that does not identify Wegener as a geophysicist, then nothing can, and we must all
retire to Bedlam.

wedded to the breakdown and foundering of continents, and Americans dedicated to the idea of continental permanence (Wegener 1912a:278). This situation was made more serious by the fact that those Americans who had developed the idea of isostasy had allied to the doctrine of continental permanence the doctrine of *oceanic* permanence, which Wegener considered to be extremely doubtful. Both sides in the debate had legitimate and even compelling evidence for their views, and yet both drew unacceptable inferences from them. Wegener offered the theory of continental drift as a working hypothesis that might reconcile the legitimate claims of both groups (Wegener 1912b:187).

Wegener understood the historical role of his thesis of continental drift; he began "Die Entstehung der Kontinente" with a history of the idea and the reasons he had proposed it. He saw its beginning in the English theory of the partition of the moon from the earth, especially in the work of Osmond Fisher (who, he said, had got the idea from G. H. Darwin and Henri Poincaré). Pickering's theory of continental drift was taken from the ideas of Fisher and Darwin and included the separation of North America from Europe-Africa (part of Wegener's own theory). However, as a result of the criticisms advanced by Schwartzchild and Liapunov of the theory of the partition of the moon, Wegener dismissed Pickering's work, "enmeshed in an incorrect hypothesis," as of historical interest only (Wegener 1912b:185).

The work of Taylor merited more serious consideration and represented a real challenge to any claims of priority that Wegener might have entertained. He was therefore quick to assert that his work and Taylor's were completely independent. Taylor, unlike Pickering, had postulated continental drift in geologically well-known epochs and had correlated it with the great Tertiary fold system of mountains. His argument for the pulling away of Greenland from North America was based on the parallelism of the coasts of the Atlantic continents. All of these were important points for Wegener's own hypothesis. A serious objection to Taylor's work was Wegener's well-founded charge that Taylor had not understood the immense consequences associated with the horizontal displacement of continents and that he had offered no discussion of the relative possibility of such an occurrence. This lacuna was so great that Wegener felt Taylor's work could only be greeted with a "shake of the head" (Wegener 1912b:185).

Taylor had at least understood the hopeless situation of the contraction theory. Geophysical evidence had been piling up against

Suess and Heim's old theory of the folding of the crust in accommodation to a shrinking interior and the idea of the creation of the continents and oceans in their present form by the sinking of continental blocks. The once-sacrosanct theory that the earth must be cooling was now challenged by the theory of radioactivity: in fact, the earth's internal temperature might even be *rising*. An independent and equally compelling argument against the contraction theory and associated continental foundering was the demonstration that the continents and oceans differed so in density that the continents were actually compensated isostatically.

Wegener reviewed the history of the isostatic concept from the time of Pratt and Airy, discussing the relative merits of the views of Hecker, Helmert, Faye, Dutton, Willis, Hayford, and J. Lukashevich on the various proper schemes for measuring gravity and reducing observations to normal values. He related these to the seismological studies of Emil Weichert and Andrij Mohorovičić and the theory of earth shells that they implied (Wegener 1912b: 187–188). He concluded that Hecker had demonstrated that the oceans were more dense than the continents and that Lukashevich had demonstrated that Airy's isostasy theory was more likely than Pratt's version of compensation. He took issue with Hayford's assumption that local details of relief were compensated but agreed with him that the continents were floated, like icebergs, on a more dense substratum.[2]

The notion that the continents floated on a dense substratum was not evidence of drift, of course; on the contrary it had been urged as an evidence for the permanence of continents and oceans. However, a completely different body of geological and paleontological evidence argued that the continents were once continuous. Structural, stratigraphical, petrological, and paleontological data argued overwhelmingly that there was an unhindered communication between the various continents at various points in the past,

2. If Wegener was misled in believing the continents to be floating, as of 1912 he had contemporary expertise on his side. Hayford had written in 1909, "The continents will be floated, so to speak, because they are composed of relatively light material; and similarly, the floor of the ocean will, on this supposed earth, be depressed because it is composed of unusually dense material. This particular condition of approximate equilibrium has been given the name isostasy." (Hayford 1909: 66). Moreover, the Russians, who have developed a history of geology of their own with a Russian—Lomonosov, Tetyayev, Schmidt, and Lukashevich—present at every step in the advance of earth science, have history on their side this time. Wegener explicitly mentions the views of Lukashevich in his reasoning for continental drift. See Lyustikh 1960:33–34.

and this made the doctrine of oceanic permanence fully as unlikely as the doctrine of earth contraction.

The geological evidence on which Wegener based his contention was none other than that accumulated by Suess, Bertrand, Heim, and the other European defenders of the no longer defensible contraction theory. Wegener employed all this evidence as well as other aspects of the older geological synthesis to oppose the doctrine of oceanic permanence and to support continental drift. He accepted Suess's division of the earth into Sal, Sima, and Nife. He took Heim's recent upward revision of the amount of crucial shortening in the Alps from 150 to between 600 and 1,200 km as a strong argument in favor of drift over the older theory: if the compression of the Himalayas were concordantly increased, there could be no question of a sunken paleocontinent in the Southern Hemisphere—the width of the palinspastically restored Himalayas and the Indian subcontinent left no room for a foundered Gondwanaland (Wegener 1912b:193).

He was particularly attuned to the hypothesis of transatlantic continuity. At every point there was structural parallelism—even a quick look at a map showed that were there mountains in the east (Europe) there were mountains in the west (Americas) as well. The correspondences between mountains of Greenland and Scandinavia and of North America and Northern Europe, the parallelism of the Caribbean and the Mediterranean and of the South American and African tablelands were accompanied by more detailed correspondences. The gneisses of the Lofoten Islands and Greenland, the transatlantic continuity of the Carboniferous mountains established by Bertrand and Suess (the Armoricans), the continuity of the carboniferous deposits of Europe and North America, all argued for a former continuity, as did the petrologic continuities of the Canary Islands and the Antilles and that between the sediments south of the delta of the Amazon and those in the Bay of Biafra. However fragmentary the evidence, which needed further confirmation of the extent of the actual continuous formations, Wegener agreed with Suess that the correspondences were "extraordinary" (Wegener 1912b:254). Wegener's geological argument was succinct: "Wherever the sinking of former continental fragments into oceanic depths has been proposed, we will substitute the splitting and relative motion of continental floes [*Kontinentalschollen*]" (Wegener 1912b:185).

His geophysical argument in defense of this thesis was a new interpretation of the views of Fisher, Hayford, and others, combined with the theory of earth shells proposed by Suess. He argued that the continents composed of Sal had deep roots in the Sima. These extended down to a depth of 100 km and were prevented from dispersing in the Sima by their higher melting point. At a depth of 100 km the continental Sal was solid, whereas the bordering Sima was plastic. This allowed the continental bergs (*Kontinentalschollen*) to plow through the denser subcrust in the course of their drift. The density differences were very slight—2.8 for the Sal, 2.9 for the Sima—but they were crucial. Even as the continents plowed through the Sima they forced it up and incorporated it in the general deformation of the leading continental margins. Thus the moving continents were gradually growing, elevating, and enriching themselves at the expense of the crust (*Erdoberfläsche*). The history of the earth was one of increasing continental growth and elevation through time. The hypsometric curve, which showed two preferential distributions of continental and oceanic elevation, was important evidence but it was also a temporary state of affairs. The difference in level would increase through time. Where once the whole earth had been covered by a "Panthalassa"—a universal ocean (one cannot suppress a remembrance of Werner's universal ocean)—the continents in their movement had progressively increased their height over the oceans. In the future, the difference would be even greater (Wegener 1912b:191–194).

Many geological puzzles were explained by the theory of continental drift. The Andean Cordillera, which Suess had not been able to correlate with his notion of overthrust of foreland subsidence, was a consequence of the resistance of oceanic Sima against continental Sal. The difference of the Atlantic and Pacific provinces, left as a great unsolved question by Suess, was explained by the rifting apart of the continents at the Atlantic margins—a phenomenon that also explained the difference between the Atlantic and Pacific coasts.

Wegener added other evidence in favor of his theory—the distribution of the *Glossopteris* flora and the indications of a Permian glaciation on widely scattered continents of the Southern Hemisphere being the most important. He also argued that various interpretations of the displacement of the poles could be reconciled with continental drift. These arguments were important in subsequent revisions and extensions of his thesis, more so than in 1912, when he was among the few earth scientists in the world able to appreciate their significance.

In the present context, it is important to recognize that Wegener's hypothesis with its proposed evidence was aimed at two audiences—the European geologists, whom he hoped to wean from the idea of continental subsidence, and the American and English geologists, whom he hoped to dissuade from the doctrine of oceanic permanence. His presentation spoke to the concerns of both groups, and he was extraordinarily careful to characterize his idea as a working hypothesis. While recognizing that the question of the force that might cause the horizontal displacement of continents was so obvious that it could not be ignored, he felt that a postulation of any causal mechanism was premature and that the immediate question was the precise determination of the reality and nature of the displacement. The alternatives—streaming (*Strömung*) of the Sima, rotational forces, the combination of rotation and pole-fleeing (*Polflucht*)—were all reasonable, but the time was not ripe for an explicit statement of the answer to this pressing question. He was anxious for a geodetic determination of the displacement, for successive measurements of astronomically determined positions of the fastest moving portions of the crust, like Greenland.

Today Wegener's hypothesis of continental drift is celebrated as the brilliant prefiguration of the theory of plate tectonics, the hypothesis that currently rules geological theory. In 1912 it was a legitimate but very tentative deduction from a great body of geological and geophysical evidence assembled in the last quarter of the nineteenth century, one of many different hypotheses created from the same materials. It had no particular claim to predominance over the theories of Willis, Chamberlin, and Joly, and it faced similar objections.

In 1948, C. E. Wegmann wrote a monograph in which he proposed possible geological tests of the theory of continental drift. In his discussion of Wegener's work he commented parenthetically, "The ways in which Wegener's aforementioned hypothesis and Charles Darwin's theory developed are very different. While Darwin gathered arguments during a period of about twenty years until his ideas had matured, Wegener put forward a suggestion which afterwards underwent many alterations. Possibly a dissertation may someday be written about all these alterations. Many of them are highly interesting if one knows the background" (Wegmann 1948:7).

That dissertation has yet to be written. In the meantime, the very proposal of the theory of continental drift in 1912 becomes "highly interesting if one knows the background." Although highly original in content, in form it was part of the mainstream of geotec-

tonic thought. Wegener was extraordinarily well read in the literature of geology and geophysics and used the same sources, addressed the same problems, employed the same concepts, and
sought the same goal as the men whose work fills the preceding
chapters of this book.

Yet it would strike a false note to end with a consideration of
Wegener, as if the history I have attempted somehow issued in or
culminated with the proposal of the theory of continental drift.
Quite the contrary. Geology in 1912 was as fragmented, from a
theoretical point of view, as at any time since its eighteenth-century
beginnings. Geology, while a mature science, functioned without a
universally accepted theoretical framework for decades thereafter.
While Wegener developed and strengthened his hypothesis of continental drift, other workers in Europe, Asia, North America, and
Africa developed global tectonic hypotheses comparable in scope
and precision that deserve equal historical attention. The competition of these hypotheses through the next fifty years (1920–1970)
has only recently abated, and with the general acceptance of the
theory of plate tectonics, geology enjoys a degree of unification it
has not experienced since the latter part of the nineteenth century.

Epilogue

ROBERT GRAVES once offered a simple recipe for the composition of a historical work: one should begin at the beginning
and, upon reaching the end, stop. But the end of what? Geology
did not end in 1912 nor did geotectonics. Even the contraction
theory, sadly battered, found distinguished advocates for decades
thereafter. Most of the principal figures whose careers and ideas
fill the later chapters of this work were active into the 1920s and
1930s, and Bailey Willis's unquenchable zeal for geological controversy persisted until his death in 1949. The various classes of
geotectonic theory that I have treated—thermal, isostatic, purely
tectonic, drift—were propounded again and again, and were joined
by new classes—oscillation or undation theories, theories of earth
expansion, and a host of eclectic stews more distinguished by their
breadth than their cogency. The year 1912 was not triumphant,
nor should the appearance of Wegener's papers, given the current
success of the theory of continental displacements, lead one into a
specious analogy with Einstein's work of 1905 and thus to the postulation of a scientific revolution in geology. If one took place, it
was a secret well kept from geologists.

But the end of the first decade of the twentieth century *is* a
watershed in geology and geological theory. The controversies that
I have chronicled flow backward toward their foundations in the
geology of the early nineteenth century and in subject matter and
scope form a continuous and related series that culminated shortly
before the First World War.

Every major controversy had but a single outcome: geology was a larger and more difficult enterprise than had previously been imagined. Unlike those disciplines that in their best moments seem to converge toward fundamentals, geology was constantly faced with dispersive tasks. Once the nature of geologic agencies was established, geologists were forced to admit that the way in which rocks were formed was only half the puzzle and that the nature of deformation had to be considered. After decades spent in the belief that the study of mountain ranges might reveal both the nature and the pattern of deformation, it became evident that the trend and character of mountain ranges was determined by larger and more deep-seated deformations best treated at the level of continents and oceans. Finally, and with much reluctance, it was admitted that even if the histories of erosion and deposition, of the creation and destruction of relief and structure, and of life could be sketched and the pattern of continental formation and destruction understood, the dynamic agency that linked these phenomena, and its rhythm, had also to be explicated.

Having finally found a way to unify these successive efforts at generalization, geologists found that their unification was unsound and had to be abandoned. But great failures can also be great successes. The fall of the contraction theory was contemporaneous with the agreement, still respected, that geotectonic theory is the ultimate basis of correlation and the ultimate level of generality.

Geologists struggled throughout the latter half of the nineteenth century to determine the extent of their responsibility to general theory in their attack on specific problems. They were forced inexorably toward higher and higher levels of generality and toward the realization that if fieldwork dominated by a single theory was dangerous, fieldwork uninformed by any theory was impossible. The stance of pure empiricism, defended by Greenough and Hall, and the stance of limited theoretical responsibility, advanced by Reade, Dutton, and others, served more to determine the fate of these men's own ideas than to sway the community at large.

The idea that geology should proceed by calm consideration of data that would suggest the proper course of their own generalization was progressively and emphatically replaced by an admission that it is adversary proceedings that produce convictions. Communion with nature is a prelude to battle with naturalists. Geology is, to adapt the phrase of the historian E. H. Carr, a hard core of interpretation surrounded by a pulp of disputable facts. This is a hard doctrine and was adopted with great reluctance and commendable fortitude by the generation that saw the decline and fall of the beautiful synthesis erected on the basis of the contraction theory and that realized the magnitude of the task still ahead.

If in the succeeding decades geologists were divided on the content of geological theory and often operated without great confidence in the framework, they never again relented on the nature and scope of the problem. As Helene Metzger suggested in 1918, the critical event in any science is not the agreement on the answer but the agreement on the question. By 1912, and thereafter, the question was geotectonic.

This history is far from complete. Important areas of geological research—paleontology, petrology, geochemistry, the experimental simulation of deformation, seismology, and rheology, all of which bear on the formulation and solution of geotectonic questions—have received no treatment whatever. My choice of 1800 as a starting point reflects only the emergence of a continuous and coherent research community at about that time. Historical work on the eighteenth and late seventeenth centuries has shown that many of the problems that occupied Hutton and Werner had already received sophisticated treatment. I might legitimately have begun with a comparison of the views of Antonio Lazzaro Moro and Benoît de Maillet or even further back.

If the year 1912 has, as I have suggested, a symbolic significance as a time when geologists were prepared to agree on the questions that a general theory would answer, few agreed that an adequate theory existed even in outline. Those who did, apart from the theorists themselves, were generally unaware of the full range of problems. In any case, the history of modern geotectonics, which begins at this point, is still unwritten.

While several treatments exist, written "backward" from the theory of plate tectonics, these necessarily treat in a summary fashion lines of evidence that do not lead directly toward elements of the modern synthesis. The further development of the theory of isostasy, the evolution of solid-earth geophysics, the history of seismology, the robust and promising development of the theory of an expanding earth, the independent evolution of tectonics in the Soviet Union, the contributions of the Dutch school to the theory of island arcs and mobile tectonic zones, the further evolution of the purely tectonic school in Europe, the Scandinavian reformulation of metamorphic petrology, the ambitious paleogeographies of North American geology, the development of geochemistry, the full variety of theories of continental motion, the role of laboratory simulation of geodynamic processes—all these remain to be explored, and each deserves a monograph of its own. The rapid proliferation and great sophistication of subspecialities in the twentieth century puts a general history of contemporary geology, for the time being, far beyond the capacities of a single investigator and the limits of a single volume.

A final note: readers interested in the philosophy of science may (and I think ought) to conclude that the history of geological theory as I have reported it substantiates neither the older view of steady, positive progress nor any existing model of scientific revolutions. The history of geology, like the history of the earth, follows a comprehensible and explicable pattern but changes in different ways, at different times, and with different intensities, in a manner that hypotheses of linear progress or alternating stages can only caricature. To my knowledge, no one has yet studied the possibility that a science might advance vigorously in the face of general and even cheerful agreement that no adequate comprehensive theory exists. Models of "scientific rationality," developed in the clear high air of theoretical physics, seem particularly prone to asphyxiation when exposed to the atmosphere of early twentieth-century geology.

Glossary

actualism (from French *actuel*, "present"). The hypothesis that the earth has always changed in the same manner and by the same agencies as today, without the restriction on intensity of action implied by uniformitarianism (q.v.).

ancient causes. Agencies of geologic change active in earlier periods of earth history and either no longer active or impossible under present conditions.

anticlinal. Shaped in the form of an arch.

basalt. A fine-grained igneous rock (q.v.), generally found in the form of lava flows and comprising more than 90 percent of all volcanic rocks. Columnar basalts, in which basalt slowly cools as hexagonal columns perpendicular to the plane of flow, are found in many locations in Europe and North America. Historically important examples appear in the Auvergne of France and in Ireland.

baseleveling. Roughly, reduction of continental elevation by erosion to sea level. Attainment of baselevel is the end product of an erosion cycle, and a necessary prelude, in the theory of continental uplift and sinking, to marine transgression.

belemnite. A fossil mollusc of the class Cephalopoda, often conical in form and with an internally chambered shell.

caloric (obs.). The hypothetical fluid of heat. Cooling, in the caloric theory, is attributed to the loss of a weightless, hot fluid that permeated the body in question.

central massive. Historically, the plutonic or granitic mass of crystalline rock that is upthrust along the axis of a mountain range and is responsible for both the uplift and deformation of sedimentary sequences.

[295]

diastrophism. Major deformations of the earth's crust resulting in mountain ranges and geosynclines and in the creation and destruction of continents and ocean basins.

dike. An igneous (q.v.) intrusion that cuts across the plane of the sedimentary beds or structures that it invades.

dip. The angle between the principal plane of a structure and the horizon, measured perpendicular to the strike (q.v.).

downfold, downflexure, downwarp. Elastic compression or deformation of a body of rock that depresses it below some local normal surface.

elevation crater (or crater of elevation). Theory (attributed to Leopold von Buch) that the cones of volcanoes are the result of radial uplift of sedimentary strata by plutonic injection, which occasionally breaks through as an eruption. The central massives (q.v.) of a mountain chain are the result, in his theory, of a line of such craters of elevation.

epeirogenic uplift. The coherent vertical uplift of a continental platform, usually without deformation, as distinct from an orogeny (q.v.).

exposure. A geographic feature, natural or artificial, that reveals rocks and exposes them to study; for example, road cuts, stream gorges, mine shafts, and cliff faces.

facies. An assemblage of rocks formed under similar conditions, deduced from the total ensemble of characteristics of the assemblage.

fissure. Crack, crevice, or declivity. In neptunian geology, a necessary precondition for a mineral vein to form, as the fissure is filled by minerals in solution, which subsequently harden.

flysch. In Swiss geology, sediments created out of the erosion products of an Alpine mountain system in the process of formation. Sediments are formed from the detritus of emerging folds and are deformed as folding continues.

foredeep. In classic Alpine tectonics when a segment of crust is folded by lateral pressure, massive downfaulting on one side of the range creates a foredeep and determines the direction of overfolding and the transverse motion of the range as a whole "in the direction of the subsidence."

foundering. In the contraction theory of the earth, the breakdown and sinking of large segments of the earth's crust as it attempts to adjust to the diminished radius of the interior.

geanticlinal. In the geological theory of James Dana, the crust's response to lateral pressure may be the formation of a broad arch, on a continental scale, or a geosynclinal (q.v.).

geomorphology. The study of land forms.

geosynclinal. In the geological theory of James Dana the earth's crust may respond to the lateral pressures of the contracting earth by a broad downfold (q.v.) called a geosynclinal.

horizon. The location (in time) of an assemblage of rocks, based on their petrology or fossil content, or both.

horst. A residual ridge or platform produced by the downfaulting of adjacent areas on both sides (in profile).

igneous rock. Historically, a rock believed to have consolidated from molten silicates (magma). If on the surface, it is volcanic; if beneath the surface, plutonic.

induration. Hardening.

intrusion. Invasion, by igneous rock, of the host or country rock.

isoclinal packing. Tight, symmetrical, often compressive folds in which successive folds are parallel.

isogeotherm. Line of equal temperature within the earth.

littoral. Shoreline between extreme high and extreme low water; intertidal zone.

loxodromism. In the geology of Alexander von Humboldt, the preferential geographic orientation of mountain ranges and/or sedimentary assemblages.

macula. In the geology of Eduard Suess, a cavity formed between the strong crust and the shrinking interior of the contracting earth, which fills with magma in response to diminished local pressure and provides the fluid substratum into which continental foundering (q.v.) takes place.

magma. The molten fluid within the earth that is the parent of crystalline rocks, appearing at the surface of the earth as lava.

metamorphism. Alteration of preexisting rock in the solid state by heat, pressure, or chemical action.

neptunism. Historically, the theory that all or nearly all of the sediments in the geological column are chemical precipitates from a primeval ocean rich in dissolved minerals.

neritic. Pertaining to the continental shelf, from just offshore to the slope that descends steeply to the abyssal floor of the oceans.

orogeny. Mountain building as a distinct geodynamic process, or a period of such mountain building; for example, the "Alpine orogeny."

palinspastic. In geological cartography, a palinspastic map is a reconstruction of the appearance of a region before extensive folding and other dislocation.

plication (obs.). Isoclinal packing (q.v.).

plutonism. Historically, the theory that both the hardening and deformation of sedimentary sequences are due to the heat and motion of molten rock deep within the earth.

posthumous folding. In classic Alpine geology, folding that takes place after the orogeny (q.v.) in which the range was created.

rigifaction (obs.). Resistance to folding of a formerly mobile or pliable section of the earth's crust as it progressively thickens.

schist. Now describes a certain kind of metamorphic rock characterized by parallel arrangement of its mineral contents; historically, as in
the geology of Hutton, used to describe any massive crystalline rock:
"the primary Schistus."

section. Diagram of a region or assemblage viewed in cross-section;
structural analysis of a locale culminates in production of a section.

secular. Long term.

sill. An igneous intrusion that invades but does not disrupt or cut
across the bedding plane of the intruded sediments or the structural
plane of the invaded assemblage.

strike. Historically, the geographic orientation of the long axis of a
major structural feature of the earth's crust, particularly a mountain
range.

talus. Coarse rock debris at the foot of a slope, the product of erosion and weathering.

tangential movement or pressure. Movement or pressure expressed
along lines tangent to the earth's surface; locally, horizontal movement
or pressure.

Tethys. A narrow, transverse ocean postulated by Suess and Neumayr that once stretched from the Himalayas to the Caribbean and
included the present Mediterranean Sea. Tethys separated the now
foundered (q.v.) paleocontinents of the North and South Atlantic.

transgression. Advance of the ocean onto the continental platforms.

trend line. The geographic orientation and course of a major structural feature, which need not be rectilinear.

type-range. In late-nineteenth-century orography, a classic mountain
range or one characteristic of a certain type of structure; exemplar of
a class of similar ranges. For example, the Jura and Urals were "Appalachian-type" ranges.

unconformity. Any hiatus, whether temporal or structural, between
successive levels in an exposure (q.v.).

uniformitarianism. A philosophy of geology proposed by Charles
Lyell that argued that the earth today is exemplary of the rate, nature,
and intensity of geologic change throughout the history of the earth,
and that the present is absolutely the key to the past.

vitreous. Glassy.

whin. Obsolete name for basalt.

Bibliography

ADAMS, FRANK D. 1912. "An Experimental Contribution to the Question of the Depth of the Zone of Flow in the Earth's Crust." *Journal of Geology*, 29:97–118.

———. 1938. *The Birth and Development of the Geological Sciences*. Baltimore: Williams and Wilkins. (Reprinted 1954. New York: Dover.)

ADHÉMAR, J. 1842. *Révolutions de la mer*. Paris.

AGER, DEREK V. 1973. *The Nature of the Stratigraphical Record*. New York: Wiley.

AIRY, SIR GEORGE B. 1855. "On the Computation of the Effect of Mountain-masses as Disturbing the Apparent Astronomical Latitude of Stations in Geodetic Surveys." *Philosophical Transactions of the Royal Society of London*. 101–104.

ALDRICH, MICHELLE L. 1976. "Frank Bursley Taylor 1860–1938." *Dictionary of Scientific Biography*, XIII: 269–272.

AMPFERER, OTTO. 1906. "Über das Bewegungsbild von Faltengebirgen." *Jahrbuch der k.k. geol. Reichsanstalt*, 56:538–622.

ANDERSON, CHARLES. 1809. See Werner 1791.

ANON. 1816. "Lectures géologiques par M. de la Métherie." *Bibliothèque Universelle des Sciences et des Arts de Genève*, 3:30.

ANON. 1817. "Memoir of Abraham Gottlob Werner, Late Professor of Mineralogy at Freiberg." *Philosophical Magazine*, 50:182–189.

ANON. 1825. "Sur la constance du niveau des mers en général et de la mer Baltique en particulier." *Bibliothèque Universelle des Sciences et des Arts de Genève*, 2:201–207.

ANON. 1935. "John Joly 1857–1933." *Obituary Notices of the Royal Society 1932–1935*. Vol. I. London: Harrison & Sons.

ARGAND, ÉMILE. 1922. "La tectonique de l'Asie." *C.R. XIII Cong. Int. de Géologie, Bruxelles 1922*. Liège, 1924. 171–372.

AUBOUIN, JEAN. 1960. *Geosynclines*. Amsterdam: Elsevier.

AVEBURY (LORD). 1896. *The Scenery of Switzerland*. London: Macmillan.

BABBAGE, CHARLES. 1837. "On the Action of Existing Causes in Producing Elevations and Subsidences in Portions of the Earth's Crust." *The Ninth Bridgewater Treatise: A Fragment*. Appendix G. London: John Murray.

BAILEY, E. B. 1935. *Tectonic Essays, Mainly Alpine*. Oxford: Clarendon.

——. 1952. *The Geological Survey of Great Britain*. London: Thos. Murby.

——. 1962. *Charles Lyell*. New York: Doubleday.

——. 1967. *James Hutton: The Founder of Modern Geology*. Amsterdam: Elsevier.

BALTZER, A. 1873. *Der Glaernisch: Ein Problem Alpiner Gebirgsbaus*. Zurich.

BARRELL, JOSEPH ("J. B."). 1910. "*Das Antlitz der Erde*." *American Journal of Science*, 179:269–270.

——. 1919. "The Nature and Bearing of Isostasy." *American Journal of Science*, 4th series. 48:281–290.

BAUMGARTEL, HANS. 1969. "Alexander von Humboldt: Remarks on the Meaning of Hypothesis in His Geological Researches." In C. J. Schneer, ed., *Toward a History of Geology*. Cambridge, Mass.: MIT Press.

BEMMELEN, RIJN VAN. 1961. "The Scientific Character of Geology." *Journal of Geology*, 69:454–463.

BERGMAN, TORBERN. 1769. *Physicalische Beschreibung der Erdkugel*. Greifswald.

BERINGER, C. C. 1954. *Geschichte der Geologie und des geologischen Weltbildes*. Stuttgart: Ferdinand Enke.

BERTRAND, MARCEL. 1884. "Rapports de structure des Alpes de Glaris et du bassin houiller du Nord." *Bulletin de la Société Géologique de France*, 3d series. 12:318–330.

——. 1887. "La chaîne des Alpes et la formation du continent européen." *Bulletin de la Société Géologique de France*, 3d series. 15:423–447.

——. 1892. "Physique du globe—Sur la déformation de l'écorce terrestre." *C. R. Acad. Sci. Paris*, 94:402–406.

BIREMBAUT, ARTHUR. 1970. "Ami Boué 1794–1881." *Dictionary of Scientific Biography*, II:341–342.

——. 1971. "Jean Baptiste Armand Louis Léonce Elie de Beaumont 1798–1874." *Dictionary of Scientific Biography*, IV:347–350.

BITTNER, A. 1881. "Ueber die geologischen Aufnamen in Judicarien und val Sabbia." *Jahrbuch der k.k. geol. Reichsanstalt*, 31:219ff.

BOUÉ, AMI. 1834. "On the Theory of Elevation of Mountain Chains, as Advocated by M. Elie de Beaumont." *Edinburgh New Philosophical Journal*, 17:123–150.

BOWIE, WILLIAM. 1917. *Investigations of Gravity and Isostasy*. U.S. Coast and Geodetic Survey, Special Publication 40.

BREISLAK, SCIPIONE. 1818. *Institutiones geologiques*. 3 vols. Milan.

BRINKEMANN, ROLAND. 1960. *The Geological Evolution of Europe*. New York: Hafner.

BROCCHI, GIOVANNI. 1814. *Conchiologia fossile subapennina con osservationi geologiche sugli Apennini e sul suolo adiacente*. Milan.

BUCH, LEOPOLD VON. 1813. *Travels through Norway and Lapland during the*

Years 1806, 1807, and 1808. Trans. John Black; notes and additional material by Robert Jameson. London.

BULLEN, K. E. 1974. *The Earth's Density.* New York: Wiley.

BURCHFIELD, JOE D. 1975. *Lord Kelvin and the Age of the Earth.* New York: Science History Publications.

BURKE, JOHN G. 1966. *Origin of the Science of Crystals.* Berkeley: University of California Press.

——. 1971. "Pierre-Louis Antoine Cordier 1777–1861." *Dictionary of Scientific Biography,* III:411–412.

CALLAWAY, CHARLES. 1883. "Origins of the Newer Gneissic Rocks of the Northern Highlands." *Quarterly Journal of the Geological Society of London,* 39:355–414.

CANNON, WALTER F. 1960. "The Uniformitarian-Catastrophist Debate." *Isis,* 51:38–55.

CAROZZI, A. V. 1972. "Emil Haug 1861–1927." *Dictionary of Scientific Biography,* VI:168–169.

CASSIRER, ERNST. 1950. *The Problem of Knowledge: Philosophy, Science, and History since Hegel.* New Haven: Yale University Press.

CHAMBERLIN, THOMAS C. 1898a. "The Ulterior Basis of Time Divisions and the Classification of Geologic History." *Journal of Geology,* 6: 450ff.

——. 1898b. "A Systematic Source for the Evolution of Provincial Faunas." *Journal of Geology,* 6:599ff.

——. 1910. "Diastrophism as the Ultimate Basis of Correlation." In Bailey Willis and Rollin D. Salisbury, (eds.), *Outlines of Geologic History with Especial Reference to North America.* Chicago: University of Chicago Press. 298–306.

CHAMBERLIN, THOMAS C., and ROLLIN D. SALISBURY. 1905–1907. *Geology.* 3 vols. 2d ed. New York: Holt.

CHAMBERLIN, THOMAS C., et al. 1917. "Geology." In *Science and Learning in France with a Survey of Opportunities for American Students in French Universities.* Society for American Fellowships in French Universities. 115–121.

CHORLEY, R. J. 1963. "The Diastrophic Background to Twentieth Century Geomorphological Thought." *Geological Society of America, Bulletin,* 74: 953–970.

CHORLEY, R. J., ANTHONY J. DUNN, and ROBERT P. BECKINSALE. 1964. *The History of the Study of Landforms, or the Development of Geomorphology.* Vol. I. London: Methuen.

——. 1973. *The History of the Study of Landforms, or the Development of Geomorphology.* Vol. II. London: Methuen.

CLOOS, HANS. 1928. *Bau und Bewegung der Gebirge.* Berlin: Borntrager.

——. 1936. *Einfuhrung in die Geologie, Ein Lehrbuch der inneren Dynamik.* Berlin: Borntrager.

——. 1947. *Gesprach mit der Erde. Geologische Welt- und Lebensfahrt.* Munich: Piper.

——. 1953. *Conversations with the Earth.* New York: Alfred Knopf.

COLLET, LEON. 1928. *The Structure of the Alps.* London: Edward Arnold.

——. ("L. W. C."). 1938. "Obituary Notice of Albert Heim." *Proceedings of the Geological Society of London,* 94:cxi–cxiii.

CORDIER, P. 1827. "Essai sur la température de l'intérieur de la terre." *Mémoires de l'Académie des Sciences pour l'année 1827,* 7:473–556.

——. 1828. "Essai sur la température de l'intérieur de la terre." *Bibliothèque Universelle des Sciences et des Arts de Genève,* 1:85–118.

CUVIER, GEORGES. 1819. "Memoir of Werner." (Reprinted 1860 in *The Naturalist's Library,* 29:19–40, ed. Wm. Jardine. London: Bohn. All page references are to the 1860 edition.)

——. 1825. *Discours sur les révolutions de la surface du globe, et sur les changements qu'elles ont produits dans le règne animal.* Paris.

DANA, JAMES D. 1847a. "On the Origin of Continents." *American Journal of Science,* 2d series. 7:94–100.

——. 1847b. "Origin of the Grand Outline Features of the Earth's Crust." *American Journal of Science,* 2d series. 9:381–398.

——. 1849. "On the Nature of Volcanic Eruptions." *United States Exploring Expedition during the Years 1838, 1839, 1840, 1841, 1842, under the Command of Charles Wilkes, U.S.N.* Vol. X. Philadelphia.

——. 1873a. "On the Origin of Mountains." *American Journal of Science,* 3d series. 5:347–350.

——. 1873b. "On Some Results of the Earth's Contraction from Cooling, Including a Discussion of the Origin of Mountains, and the Nature of the Earth's Interior." *American Journal of Science,* 3d series. 5:423–443.

DARWIN, CHARLES. 1882. *The Autobiography of Charles Darwin 1809–1882.* London. (Reprinted 1958. London: Collins. All page references are to the 1958 edition.)

D'AUBUISSON DES VOISINS, J. F. 1819. *Traité de géognosie.* Paris: F. G. Levrault.

DAVIES, GORDON L. 1969. *The Earth in Decay: A History of British Geomorphology 1578–1878.* New York: American Elsevier.

DAVIS, W. M. 1919. "The Framework of the Earth." *American Journal of Science,* 48:225–241.

DE LA BECHE, HENRY T. 1834. *Researches in Theoretical Geology.* London: G. Knight.

——. 1835. *A Geological Manual.* 3d. ed. London.

——. 1846. "On the Formation of South Wales and South-Western England." *Memoirs of the Geological Survey of Great Britain,* I:221ff.

DE LAUNAY, L. 1913. *La science géologique: Ses méthodes, ses résultats, ses problèmes, son histoire.* Paris: Colin.

DELUC, J. A. 1809. *An Elementary Treatise on Geology.* London.

——. 1820. "Considérations sur la stabilité des montagnes." *Bibliothèque Universelle des Sciences et des Arts de Genève,* 3:143–147.

——. 1821. "De la chaleur intérieure de la terre." *Bibliothèque Universelle des Sciences et des Arts de Genève,* 3:40–61.

DE MARGERIE, EMMANUEL. 1946. "Three Stages in the Evolution of Alpine Geology: DeSaussure-Studer-Heim." *Quarterly Journal of the Geological Society of London,* 102:xcvii–cxiv.

DE MARGERIE, EMMANUEL, and ALBERT HEIM. 1888. *Les dislocations de l'écorce terrestre. Die Dislocationen der Erdrinde. Essai de définition et de nomenclature. Versuch einer Definition und Bezeichung.* Zurich: J. Wurster.

DUTTON, CLARENCE E. 1883. "Physics of the Earth's Crust." *American Journal of Science,* 123:283–290.

——. 1889. "On Some Greater Problems of Physical Geology." (Reprinted 1925 in *Journal of the Washington Academy of Sciences,* 15:359–369. All page references are to the 1925 publication.)

ELIE DE BEAUMONT, L. 1824. *Coup d'oeil sur les mines.* Paris: F. G. Levrault.

——. 1828. *Observations géologiques sur les différents formations qui, dans le système des Vosges, séparent la formation houillier de celle du lias.* Paris.

——. 1831. "Researches on Some of the Revolutions on the Surface of the Globe; Presenting Various Examples of the Coincidence between the Elevation of Beds in Certain Systems of Mountains, and the Sudden Changes Which Have Produced the Lines of Demarcation Observable in Certain Stages of the Sedimentary Deposits." *Philosophical Magazine,* n.s. 10:241–264.

——. 1833. "Sur quelques points de la question des cratères de soulève-ment." *Bulletin de la Société Géologique de France,* 4:225–291.

——. 1844. "Note sur la rapport qui existe entre le réfroidissement pro-gressif de la masse du globe terrestre et celui de la surface." *C. R. Acad. Sci. Paris,* 19:1327–1331.

——. 1850. "Note sur la corrélation des directions des différents systèmes de montagnes." *C. R. Acad. Sci. Paris,* 31:325–338.

——. 1852. *Notice sur les systèmes des montagnes.* 3 vols. Paris: P. Bertrand. (The pages of the three volumes are numbered consecutively.)

ESCHER, J. C. 1820. "Rapports de la troisième séance, Société Helvétique d'Histoire Naturelle." *Bibliothèque Universelle des Sciences et des Arts de Genève,* 3:75.

EYLES, V. A. 1964. "Abraham Gottlob Werner (1749–1817) and His Position in the History of the Mineralogical and Geological Sciences." *History of Science,* 3:102–115.

FAVRE, ERNEST. 1883. "Revue géologique suisse pour l'année 1882." *Archives des sciences physiques et naturelles,* 9:175–220, 279–331.

FISHER, DONALD 1972. "James Hall Jr. 1811–1898." *Dictionary of Scientific Biography,* VI:56–58.

FISHER, JEROME D. 1963. "Mineralogical Society of America." *Geotimes,* 7:8–12.

FISHER, OSMOND. 1874. "On the Inequalities of the Earth's Surface Viewed in Connection with the Secular Cooling." *Geological Magazine,* 1:60–63.

——. 1876. "On the Inequalities of the Earth's Surface Viewed in Connec-tion with the Secular Cooling." *Cambridge Philosophical Society Proceedings,* 2:324–331.

——. 1878. "On the Possibility of Changes in the Latitude of Places on the Earth's Surface." *Geological Magazine,* 5:291–297, 551–552.

——. 1879. "On the Formation of Mountains on the Hypothesis of a Liquid Substratum." *Cambridge Philosophical Society Transactions,* 12:434–454.

——. 1881. *The Physics of the Earth's Crust.* London: Macmillan. (2d ed., 1889.)

——. 1882. "On the Physical Cause of the Ocean Basins." *Nature*, 25: 243–244.

——. 1883. "Physics of the Earth's Crust." *Nature*, 27:76–77.

——. 1886. "On the Variations of Gravity at Certain Stations of the Indian Arc of the Meridian in Relation to Their Bearing upon the Constitution of the Earth's Crust." *Philosophical Magazine*, 22:1–29.

——. 1887. "A reply to the Objections Raised by Mr. Charles Davison M.A. to the Argument on the Insufficiency of the Contraction of the Solid Earth to Account for the Inequalities or Elevations of the Surface." *Philosophical Magazine*, 24:391–394.

——. 1888. "On the Mean Height of Surface Elevations and Other Quantitative Results of the Contraction of a Solid Globe through Cooling; Regard Being Paid to the Existence of a Level of No Strain, as Lately Announced by Mr. T. Mellard Reade and by Mr. C. Davison." *Philosophical Magazine*, 25:7–20.

FLEURE, H. J. 1955. "Emmanuel de Margerie (1862–1953)." *Biographical Memoirs of Fellows of the Royal Society*. London: Royal Society, 184–191.

FOURIER, J. 1820. "Mémoire sur le réfroidissement séculaire du globe terrestre." *Société Philomathique, bulletin.* 81–87, 156–165.

——. 1837. "General Remarks on the Temperature of the Terrestrial Globe and Planetary Spaces." *American Journal of Science*, 32:1–20.

FOURNET, JOSEPH. 1850. "Adresse à la Société d'Agriculture et de l'Histoire Naturelle de Lyon." *Archives des Sciences Physiques et Naturelles*, 1:243–245.

FRÄNGSMAYR, TORE. 1976. "The Geological Ideas of J. J. Berzelius." *British Journal for the History of Science*, 9:228–236.

FRANKEL, HENRY. 1976. "Alfred Wegener and the Specialists." *Centaurus*, 20:305–324.

FRAZER, PERSIFOR. 1888. "Account of the First International Congresses." *American Geologist*, 1:4–10.

GEIKIE, ARCHIBALD. 1905. *The Founders of Geology*. London: Macmillan.

GERSTNER, PATSY A. 1968. "James Hutton's Theory of the Earth and His Theory of Matter." *Isis*, 59:26–31.

——. 1975. "A Dynamic Theory of Mountain-Building: Henry Darwin Rogers, 1842." *Isis*, 66:26–37.

GILBERT, G. K. 1895. "Notes on the Gravity Determinations Reported by Mr. Putnam." *Bulletin of the Philosophical Society of Washington*, 13:61–75.

GILLISPIE, CHARLES COULSTON. 1959. *Genesis and Geology*. New York: Harper. (First published in 1951.)

GOSSELET, L. (1879/1880). "Sur la structure générale du bassin houiller franco-belge." *Bulletin de la Société Géologique de France*, 3:505.

GRATTAN-GUINNESS, I. 1972. *Joseph Fourier 1768–1830*. Cambridge, Mass.: MIT Press.

GREGORY, J. W., Ed. 1929. *The Structure of Asia*. London: Methuen.

GROSS, P. L. K., and O. A. WOODFORD. 1931. "Serial Literature Used by American Geologists." *Science*, 73:660–664.

HALL, JAMES. 1805. "Experiments on Whinstone and Lava." *Transactions of the Royal Society of Edinburgh*, V:43ff.

——. 1812. "Account of a Series of Experiments." *Transactions of the Royal Society of Edinburgh*, VI:73–77, 80–88, 151–155.

——. 1815. "On the Vertical Position of Certain Strata and Their Relations with the Granite." *Transactions of the Royal Society of Edinburgh*, VII:79ff.

——. 1826. "On the Consolidation of the Strata." *Transactions of the Royal Society of Edinburgh*, X:314–329.

HALL, JAMES, JR. 1859. *Paleontology*. Vol. III. Geological Survey of New York. Albany: Van Benthuysen.

——. 1882. "Contributions to the Geological History of the North American Continent." *Proceedings of the American Association for the Advancement of Science*, 31:29–71. (Presidential address of 1857.)

HALLAM, A. 1973. *A Revolution in the Earth Sciences: From Continental Drift to Plate Tectonics*. Oxford: Clarendon Press.

HALLER, JOHN. 1976. "Pierre Termier 1859–1930." *Dictionary of Scientific Biography*, XIII: 283–286.

HAUG, ÉMILE. 1900. "Les géosynclinaux et les aires continentales: Contribution à l'étude des transgressions et régressions marines." *Bulletin de la Société Géologique de France*, 3d series. 28:617–710.

——. 1907–1911. *Traité de géologie*. 3 vols. Paris: Colin.

——. 1925. "Contribution à une synthèse des Alpes occidentales." *Bulletin de la Société Géologique de France*, 4th series. 25:97–244.

——. 1907–1911. *Traité de géologie*. 3 vols. Paris: Colin.

HAYFORD, JOHN F. 1909. *The Figure of the Earth and Isostasy from Measurements in the United States*. Washington, D.C.: Government Printing Office.

HECKER, O. 1903. "Bestimmung der Schwerkraft auf dem Atlantischen Ocean." *Veröff. k. preuss. geodät Inst.*, no. 11.

——. 1908. "Bestimmung der Schwerkraft auf dem Indischen und Grossen Ocean." *Veröff. k. preuss. geodät Inst.*, no. 12.

HEIM, ALBERT. 1878. *Untersuchungen über den Mechanismus der Gerbirgsbildung: im Anschluss an die geologische Monographie der Todi-Windgällen Gruppe*. 2 vols. and atlas. Basel: Benno Schwabe.

——. 1919–1921. *Geologie der Schweiz*. 3 vols. Leipzig: Chr. Herman Tauchnitz.

HEIMANN, P. M., and J. E. McGUIRE. 1971. "Newtonian Forces and Lockean Powers: Concepts of Matter in Eighteenth Century Thought." *Historical Studies in the Physical Sciences*. Vol. 3. Philadelphia: University of Pennsylvania Press.

HEISKANEN, W. 1924. "Die Airysche isostatische Hypothese und Schwermessung." *Zeitschrift fur Geophysik*, 1:225–227.

HERITSCH, FRANZ. 1928. *The Nappe Theory in the Alps: Alpine Tectonics 1905–1928*. Trans. P. G. H. Boswell. London: Methuen.

HERIVEL, J. 1975. *Joseph Fourier: The Man and the Physicist*. Oxford: Clarendon Press.

HERSCHEL, JOHN. 1837. "Letter to Lyell." In Charles Babbage, *The Ninth Bridgewater Treatise*. London: John Murray. 202–217.

HOBBS, WILLIAM HERBERT. 1914. "Eduard Suess." *Journal of Geology*, 22: 811–817.

——. 1919. "La face de la terre." *Journal of Geology*, 27:133–136.

HOLDER, HELMUT. 1960. *Geologie und Paläontologie in Texten und ihrer Geschichte*. Freiberg/Munich: Karl Alber.

HOLMES, ARTHUR. 1965. *Principles of Physical Geology*. 2d ed. New York: Ronald Press.

HOLTEDAHL, OLAF. 1920. "Paleogeography and Diastrophism in the Atlantic-Arctic Region during Paleozoic Time." *American Journal of Science*, 199: 1–25.

HOOYKAAS, R. 1963. *Natural Law and Divine Miracle: The Principle of Uniformity in Geology, Biology and Theology*. Leiden: E. J. Brill.

——. 1970. *Catastrophism in Geology, Its Scientific Character in Relation to Actualism and Uniformitarianism*. Amsterdam: North-Holland.

——. 1972. "Rene-Just Haüy." *Dictionary of Scientific Biography*, VI:178–183.

HUNT, T. STERRY. 1873. "On Some Points in Dynamical Geology." *American Journal of Science*, 3d series. 5:264–270.

HUTTON, JAMES. 1788. "On the Theory of the Earth; or, An Investigation of the Laws Observable in the Composition, Dissolution, and Restoration of the Globe." *Transactions of the Royal Society of Edinburgh*, I:209–304.

——. 1795. *Theory of the Earth with Proofs and Illustrations*. 2 vols. Edinburgh: Wm. Creech. (Reprinted 1960. London: Wheldon and Wesley.)

JAMESON, ROBERT. 1800. *Mineralogy of the Scottish Isles*. 2 vols. Edinburgh.

JEFFREYS, SIR HAROLD. 1970. *The Earth: Its Origin, History, and Physical Constitution*. 5th ed. Cambridge: Cambridge University Press.

JOLY, JOHN. 1909. *Radioactivity and Geology*. London: Constable.

——. 1925. *The Surface History of the Earth*. Oxford: Clarendon Press.

JONES, O. T. 1937. "On the Evolution of a Geosyncline." *Proceedings of the Geological Society of London*, 94:lx–cx.

KAY, MARSHALL, and EDWIN COLBERT. 1964. *Stratigraphy and Life History*. New York: Wiley.

KEFERSTEIN, CHRISTIAN. 1840. *Geschichte und Literatur der Geognosie*. Halle.

KENDALL, MARTHA. 1970. "Joseph Barrell 1869–1919." *Dictionary of Scientific Biography*, I:469–471.

KILIAN, W. 1891. "Notes sur l'histoire et la structure géologique des chaînes alpines." *Bulletin de la Société Géologique de France*, 19:571–661.

KITTS, DAVID B. 1973. "Grove Karl Gilbert and the Concept of 'Hypothesis' in Late Nineteenth Century Geology." In R. Geire and R. Westfall, eds., *Foundations of Scientific Method: The Nineteenth Century*. Bloomington: Indiana University Press.

——. 1974. "Physical Theory and Geological Knowledge." *Journal of Geology*, 82:259–272.

LAPWORTH, CHARLES. 1883. "On the Secret of the Highlands." *Geological Magazine*, X:120–128, 193–199.

LeCONTE, JOSEPH. 1873. "Letter to the Editor, Jan. 13, 1873." *American Journal of Science*, 5:156.

LEHMANN, J. G. 1753. *Abhandlung von den Metalmüttern und der Erzeugung der Metalle*. Berlin.

LOEWINSON-LESSING, F. Y. 1954. *A Historical Survey of Petrology*. London: Oliver and Boyd. (First published in Russian, 1938.)

LONGWELL, CHESTER R. 1941. "Geology." In L. L. Woodruff, ed., *The Development of the Sciences.* New Haven: Yale University Press. 147–196.

LUGEON, MAURICE. 1896. "La région de la brèche du Chablais." *Bull. Serv. Carte Géologique de France,* VII:49ff.

——. 1901a. "Les grands nappes de récouvrement des Alpes du Chablais et de la Suisse." *Bulletin de la Société Géologique de France,* 4th series. 1:723–825.

——. 1901b. "Sur la découverte d'une racine des Prealpes suisses." *Bulletin de la Société Géologique de France,* 4th series. 1:45ff.

LYELL, CHARLES. 1830–1833. *Principles of Geology, Being an Attempt to Explain the Former Changes of the Earth's Surface by Reference to Causes Now in Operation.* 3 vols. London: James Murray.

——. 1835. "Sur les prévues d'une élévation graduelle du sol dans certaines parties de la Suède." *Bibliothèque Universelle des Sciences et des Arts de Genève,* 2:326–328.

LYUSTIKH, E. N. 1960. *Isostasy and Isostatic Hypotheses.* Soviet Research in Geophysics, vol. 2. Trans. American Geophysical Union. New York: Consultants Bureau. (First published in Russian, 1957.)

MACMILLAN, W. D. 1917. "On the Hypothesis of Isostasy." *Journal of Geology,* 25:105–111.

MALLET, ROBERT. 1873. "Note on the History of Certain Recent Views in Dynamical Geology." *American Journal of Science,* 3d series. 5:302–303.

MANHEIM, JEROME H. 1975. "John Henry Pratt 1809–1871." *Dictionary of Scientific Biography,* XI:126–127.

MARKHAM, CLEMENTS R. 1878. *A Memoir on the Indian Surveys.* 2d ed. London: W. H. Allen.

MARVIN, URSULA. 1973. *Continental Drift: The Evolution of a Concept.* Washington, D.C.: Smithsonian Institution Press.

MATHER, KIRTLEY F. 1971. "Thomas Crowder Chamberlin 1843–1928." *Dictionary of Scientific Biography,* III:189–191.

MATHER, KIRTLEY F., and SHIRLEY MASON, eds. 1939. *A Source Book in Geology 1400–1900.* Cambridge, Mass.: Harvard University Press.

MATHEWS, EDWARD B. 1927. "Progress in Structural Geology." In Edward B. Mathews, ed., *Fifty Years Progress in Geology: 1876–1926.* Baltimore: Johns Hopkins University Press.

MCCOSH, JAMES. 1875. *The Scottish Philosophy.* New York: Carter.

MERRILL, G. P. 1924. *The First One Hundred Years of American Geology.* New Haven: Yale University Press.

METZGER, Hélène. 1918. *La genèse de la science des cristaux.* (Reprinted 1969. Paris: Blanchard.)

MEUNIER, STANISLAS. 1911. *L'évolution des théories géologiques.* Paris: Alcan.

MILLER, G. A. 1918. "Scientific Activity and the War." *Science,* 48:117–118.

MOLENGRAFF, G. A. F. 1913. "Folded Mountain Chains, Overthrust Sheets and Block-Faulted Mountains in the East Indian Archipelago." *C.R. XII Cong. Int. de Géologie.* Canada, 1913. 689–702.

MOORE, RAYMOND C. 1941. "Stratigraphy." *Geology 1888–1938. Geological Society of America, Fiftieth Anniversary Volume.* New York: Geological Society of America. 179–220.

MOULTON, FOREST RAY. 1929. "T. C. Chamberlin as a Philosopher." *Journal of Geology*, 37:368–379.

MURCHISON, RODERICK I. 1845. *Geology of Russia and the Ural Mountains*. London.

MURRAY, JOHN. 1802. *A Comparative View of the Huttonian and Neptunian Systems of Geology: In Answer to the Illustrations of the Huttonian Theory of the Earth by Professor Playfair*. Edinburgh: Ross and Blackwood.

NECKER DE SAUSSURE, L. A. 1821. *Voyage en Ecosse et aux Iles Hebrides*, 3 vols. Geneva: Paschoud.

——. 1824. "Discours sur l'histoire et les progrès de la géologie." *Bibliothèque Universelle des Sciences et des Arts de Genève*, 2:106–126.

——. 1830. "Sur quelques rapports entre la direction générale de la stratification et celle des lignes d'égale intensité magnétique dans l'hémisphere boreal." *Bibliothèque Universelle des Sciences et des Arts de Genève*, 1:1ff.

NEUMAYR, MELCHIOR. 1886. *Erdgeschichte*. Leipzig: Bibliographiches Institut. (2d ed., 1895; 3d ed., 1900.)

NICOLSON, MARJORIE HOPE. 1959. *Mountain Gloom and Mountain Glory*. Ithaca, N.Y.: Cornell University Press.

NIEUWENKAMP, W. 1970. "Leopold von Buch 1774–1853." *Dictionary of Scientific Biography*, II:552–557.

OLDROYD, D. R. 1974. "Some Phlogistic Mineralogical Schemes Illustrative of the Evolution of the Concept of 'Earth' in the 17th and 18th Centuries." *Annals of Science*, 31:269–305.

OLSON, RICHARD. 1969. "The Reception of Boscovich's Ideas in Scotland." *Isis*, 60:91–103.

——. 1975. *Scottish Philosophy and British Physics 1750–1880*. Princeton, N.J.: Princeton University Press.

OSPOVAT, ALEXANDER. 1960. "Werner's Influence on American Geology." *Proceedings of the Oklahoma Academy of Sciences*, 40:98–103.

——. 1969. "Reflections on A. G. Werner's 'Kurze Klassifikation.'" In C. J. Schneer, ed., *Toward a History of Geology*. Cambridge, Mass.: MIT Press.

——. 1976. "The Distortion of Werner in Lyell's *Principles of Geology*." *British Journal for the History of Science*, 9:190–198.

PAGE, LEROY E. 1969. "Diluvialism and Its Critics in Great Britain in the Early Nineteenth Century." In C. J. Schneer, ed., *Toward a History of Geology*. Cambridge, Mass.: MIT Press.

PAULSEN, FRIEDRICH. 1906. *The German University and University Study*. Trans. Frank Thilly. New York: Scribners. (First published in German, 1902.)

PEACH, B. N., and J. HORNE. 1884. "Report on the Geology of the North-West of Sutherland." *Nature*, 31:31ff.

PERRIER, GEORGES. 1939. *Petite histoire de la géodesie, comment l'homme a mesuré et pesé la terre*. Paris: Alcan.

PHILLIMORE, R. H. ed. 1958. *Historical Records of the Survey of India*. Vol. IV. *1830–1843: George Everest*. Dehra Dun: Survey of India.

PICKERING, WILLIAM H. 1907. "The Place of the Origin of the Moon: The Volcanic Problem." *Journal of Geology*, 15:23–38.

PLAYFAIR, JOHN. 1802. *Illustrations of the Huttonian Theory of the Earth.* Edinburgh: Wm. Creech.

PORTER, ROY. 1975a. *The Making of Geology: Earth Science in Britain 1660–1815.* New York: Cambridge University Press.

———. 1975b. "The Industrial Revolution and the Rise of the Science of Geology." In Mikulas Teich and Robert Young, eds., *Changing Perspectives in the History of Science.* London: Heinemann. 320–343.

———. 1976. "Charles Lyell and the Principles of the History of Geology." *British Journal for the History of Science,* 9:92–104.

POWELL, JOHN W. 1875. *Exploration of the Colorado River of the West and Its Tributaries. Explored in 1869, 1870, 1871, and 1872, under the Direction of the Secretary of the Smithsonian Institution.* Washington, D.C.: Government Printing Office.

PRATT, JOHN HENRY. 1855. "On the Attraction of the Himalaya Mountains, and of the Elevated Regions beyond Them, upon the Plumb-line in India." *Philosophical Transactions of the Royal Society.* 53–100.

———. 1859. "On the Deflection of the Plumbline in India, Caused by the Attraction of the Himmalaya [sic] Mountains and of the Elevated Regions Beyond; and Its Modification by the Compensating Effect of a Deficiency of Matter below the Mountain Mass." *Philosophical Transactions of the Royal Society.* 745–778.

PRÉVOST, CONSTANT. 1840. "Opinion sur le théorie des soulèvements." *Bulletin de la Société Géologique de France,* 11:183–203.

———. 1850. "Remarques à l'occasion d'un mémoire de M. Elie de Beaumont sur la corrélation des différents systèmes de montagnes." *C.R. Acad. Sci. Paris,* 31:439–444.

PUTNAM, G. R. 1895. "Results of a Transcontinental Series of Gravity Measurements." *Bulletin of the Philosophical Society of Washington,* 13:31–60.

RAUMER, KARL VON 1859. *Contributions to the History and Improvement of the German Universities.* New York: Brownell.

READ, H. H. 1957. *The Granite Controversy.* London: Thos. Murby.

READ, RALPH D. 1941. "Structural Geology." *Geology 1888–1938. Geological Society of America, Fiftieth Anniversary Volume.* New York: Geological Society of America 243–267.

READE, T. MELLARD. 1886. *The Origin of Mountain Ranges Considered Experimentally, Structurally and Dynamically, and in Relation to Their Geological History.* London: Taylor and Francis.

———. 1888a. "Elevation and Subsidence." *Geological Magazine,* 5:382–383, 432.

———. 1888b. "Mountain Formation." *Philosophical Magazine,* 25:521–522.

———. 1891. "An Outline of the Theory of the Origin of Mountain Ranges by Sedimentary Loading and Cumulative Recurrent Expansion in Answer to Recent Criticisms." *Philosophical Magazine,* 31:485–496.

———. 1889. "Physical Theories of the Earth in Relation to Mountain Formation." *American Geologist,* 3:106–111.

———. 1892a. "Physics of Mountain Building: Some Fundamental Conceptions." *American Geologist,* 9:238–243.

——. 1892b. "Causes of Deformation of the Earth's Crust." *Nature*, 46:315.

——. 1894a. "On the Results of the Unsymmetrical Cooling and Redistribution of Temperature in a Shrinking Globe as Applied to the Origin of Mountain Ranges." *Geological Magazine*, 1:203–214.

——. 1894b. "Some Physical Questions Connected with Theories of the Origin of Mountain Ranges." *Geological Magazine*, 1:413–414.

RENEVIER, E. 1879. "Notice of Prof. A. Heim's Work on the Mechanism of the Formation of Mountains." *Geological Magazine*, 6:131–135.

REYER, E. 1888. *Theoretische Geologie*. Stuttgart.

——. 1892. "On the Causes of the Deformation of the Earth's Crust." *Nature*, 46:224–227.

RICE, WILLIAM N., ed. 1915. *Problems of American Geology*. New Haven: Yale University Press.

ROGERS, HENRY DARWIN. 1858. *The Geology of Pennsylvania: A Government Survey, with a General View of the Geology of the United States*. Vol. II, pt. 2. Edinburgh, New York, Philadelphia.

ROGERS, HENRY DARWIN, and WILLIAM B. ROGERS. 1843. "On the Physical Structure of the Appalachian Chain as Exemplifying the Laws Which Have Regulated the Elevation of Great Mountain Chains Generally." *Reports of Meetings of the Association of American Geologists and Naturalists*. 474–531.

ROGERS, JOHN. 1975. "Henry Darwin Rogers and William B. Rogers." *Dictionary of Scientific Biography*, XI:504–506.

ROLLER, D. H. D. 1971. *Perspectives in the History of Science and Technology*. Norman: University of Oklahoma Press.

RUDWICK, MARTIN J. S. 1962. "Hutton and Werner Compared: George Greenough's Geological Tour of Scotland in 1805." *British Journal for the History of Science*, I:117–135.

——. 1970. "The Strategy of Lyell's *Principles of Geology*." *Isis*, 60:5–33.

——. 1971. "Uniformity and Progression." In D. H. D. Roller, ed., *Perspectives in the History of Science and Technology*. Norman: University of Oklahoma Press.

RUTTEN, M. G. 1969. *The Geology of Western Europe*. Amsterdam: Elsevier.

SAINTE-CLAIRE DEVILLE, CHARLES. 1878. *Coup d'oeil historique sur la géologie et sur les travaux d'Elie de Beaumont*. Paris: G. Masson.

SARTON, GEORGE. 1919. "La synthèse géologique de 1775 à 1918." *Isis*, 2:357–394.

SAUSSURE, H. B. de 1779–1796. *Voyages dans les Alpes*. 4 vols. Neuchâtel.

SCHARDT, HANS. 1893a. "Sur l'origine des Alpes du Chablais et du Stockhorn, en Savoie et en Suisse." *C.R. Acad. Sci. Paris*, 127:707ff.

——. 1893b. "Sur l'origine des Prealpes romandes." *Archives des sciences physiques et naturelles*, 30:570.

——. 1898. "Les régions éxotiques du versant nord des Alpes suisses." *Bulletin de la Société Vaudoise des Sciences Naturelles*, 34:113–219.

——. 1904. "Notes sur la profile géologique et la tectonique du massif du Simplon" *Eclogae Geologicae Helvetiae*, 8:173–200.

SCHEIDEGGER, ADRIAN E. 1963. *Principles of Geodynamics*. New York: Academic Press.

SCHUCHERT, CHARLES. 1929. "Chamberlin's Philosophy of Correlation." *Journal of Geology*, 37:328–340.

SEDGWICK, ADAM. 1831. "Presidential Address." *Proceedings of the Geological Society of London*, 1:311.

STEFFENS, HENRIK. 1863. *The Story of My Career*. Abridged and trans. W. L. Gage. Boston: Gould and Lincoln.

STUDER, BERNHARD. 1825. *Beytrage zu einer Monographie der Molasse*. Berne.

———. 1846. "Coup d'oeil sur la géologie des Alpes occidentales." *Archives des Sciences Physiques et Naturelles*, 3:248–264.

SUESS, EDUARD. 1875. *Die Entstehung der Alpen*. Vienna: Wilhelm Braumuller.

———. 1887. "Ueber unterbrochene Gebirgsfaltung." *Sitzungsberichte der Kaiserlich Academie der Wissenschaften*, XCIV:94.

———. 1897a. "On the Structure of Europe," *Canadian Record of Science*, 7:235–246.

———. 1897b. "Modern Attainments in Geology," *Canadian Record of Science*, 7:272–288.

———. 1901. "Farewell Address by Professor Eduard Suess on Resigning His Professorship." Trans. Charles Schuchert. *Journal of Geology*, 12 (1904):264–275.

———. 1904–1909. *The Face of the Earth*. Trans. Hertha B. C. Sollas. Vol. 1, 1904; vol. 2, 1905; vol. 3, 1906; vol. 4, 1909. Oxford: Clarendon Press. (First published in German as *Das Antlitz der Erde*. Vol. 1, 1883; vol. 2, 1888; vol. 3, 1901; vol. 4, 1904).

———. 1911. "Synthesis of the Paleogeography of North America." *American Journal of Science*, 181:101–108.

———. 1916. *Erinnerungen*. Leipzig: G. Hirzel.

Suess, F. E. 1929. "The European Altaides and Their Correlation to the Asiatic Structure." In J. W. Gregory, ed., *The Structure of Asia*. London: Methuen. 35–37.

SULLIVAN, WALTER. 1974. *Continents in Motion: The New Earth Debate*. New York: McGraw-Hill.

SWEET, J., and C. WATERSTON. 1967. "Robert Jameson's Approach to the Wernerian Theory of the Earth, 1796." *Annals of Science*, 23:81–95.

TAYLOR, FRANK B. 1910. "Bearing of the Tertiary Mountain Belt on the Origin of the Earth's Plan." *Geological Society of America, Bulletin*, 21:179–226.

TAYLOR, KENNETH L. 1969. "Nicholas Desmarest and Geology in the Eighteenth Century." In C. J. Schneer, ed., *Toward a History of Geology*. Cambridge, Mass: MIT Press.

TERMIER, PIERRE. 1899. "Les nappes de récouvrement du Brianconnais." *Bulletin de la Société Géologique de France*, 3d series. 27:47–84.

———. 1903. "Les nappes des Alpes orientales et la synthèse des Alpes," *Bulletin de la Société Géologique de France*, 4th series. 3:711–766.

———. 1908. "Eloge à Marcel Bertrand." *Bulletin de la Société Géologique de France*, 4th series. 8:163–204.

THURMANN, J. 1853/1854. "Resumé des lois orographiques du système des monts Jura." *Bulletin de la Société Géologiqiue de France*. 2d series. 11:41–50.

TIETZE, EMIL. 1916. "Einige Seiten über Eduard Suess. Ein Beitrag zur Geschichte der Geologie." *Jahrbuch de k.k. geol. Reichsanstalt*, 66:330–556.

TOBEIN, H. 1976. "Bernhard Studer 1794–1887." *Dictionary of Scientific Biography*, XIII:123–124.

VOGELSANG, H. 1867. *Philosophie der Geologie und Mikroskopische Gestein-studien*. Bonn.

VOSE, GEORGE L. 1866. *Orographic Geology; or, The Origin and Structure of Mountains: A Review*. Boston: Lee and Shepard.

WEGENER, ALFRED. 1912a. "Die Entstehung der Kontinente." *Geologische Rundschau*, 3:276–292.

———. 1912b. "Die Entstehung der Kontinente." *Petermanns Mitteilungen*, I:185–195, 253–256, 305–309.

WEGMANN, C. E. 1948. "Geological Tests of the Hypothesis of Continental Drift in the Arctic Regions." *Meddelelser om Gronland*, 144:5–49.

———. 1970. "Emile Argand, 1879–1940." *Dictionary of Scientific Biography*, I:235–237.

———. 1973. "Maurice Lugeon, 1870–1953." *Dictionary of Scientific Biography*. VIII:543–545.

———. 1976. "Eduard Suess, 1831–1914." *Dictionary of Scientific Biography*, XIII:143–148.

WERNER, A. G. 1774. *Von den äusserlichen Kennzeichen der Fossilien*. (English translation 1962 by A. V. Carozzi. *On the External Character of Minerals*. Urbana: University of Illinois. All page references are to the 1962 edition.)

———. 1785. *Von den ausserlichen Kennzeichen der Fossilien*. Vienna Edition with emendations. (Unauthorized.)

———. 1787. *Kurze Klassifikation und Beschreibung der verschiedenen Gebirgsarten*.

———. 1790. *Traité des charactères extérieurs des fossiles*. Trans. M. Picardet. Dijon. (Unauthorized.)

———. 1791. *Neue Theorie von der Entstehung der Gange, mit Anwendung auf den Bergbau, besonders den Freibergschen*. Freiberg. (English translation 1809 by Charles Anderson. *New Theory of the Formation of Veins, with Its Application to the Art of Working Mines*. Edinburgh: Encylopedia Britannica Press. All page references are to the 1809 edition.)

———. 1805. *A Treatise on the External Character of Fossils*. Trans. and emended by Thomas Weaver. Dublin. (Unauthorized.)

WHEWELL, WILLIAM. 1837. *History of the Inductive Sciences from the Earliest Times to the Present*. London: Parker.

WILLIS, BAILEY. 1893. "Mechanics of Appalachian Structure." *Annual Report of the U.S. Geological Survey 1891–1892*, 13:211–281.

———. 1906. "Geologic Research in Continental Histories." *Yearbook of the Carnegie Institution*, 4:204–214.

———. 1907a. "Thrusts and Recumbent Folds: A Suggestion Bearing on Alpine Structure." *Science*, n.s. 25:1010–1011.

———. 1907b. "A Theory of Continental Structure Applied to North America." *Geological Society of America, Bulletin*, 18:389–412.

———. 1910a. "Principles of Paleogeography." *Science*, n.s. 31:241–260.

——. 1910b. "What Is Terra Firma? A Review of Current Research on Isostasy." *Annual Report of the Smithsonian Institution.* 391–406.

——. 1912. "Report on an Investigation of the Geological Structure of the Alps." *Smithsonian Miscellaneous Collections,* 56, no. 31:1–13.

WILLIS, BAILEY, and ROBIN WILLIS. 1934. *Geologic Structures.* 3d ed. New York: McGraw-Hill.

WILSON, LEONARD. 1972. *Charles Lyell, the Years to 1841: The Revolution in Geology.* New Haven: Yale University Press.

WOODWARD, R. S. 1889. "The Mathematical Theories of the Earth." *American Journal of Science,* 38:337–355.

ZITTEL, K. A. VON. 1901. *History of Geology and Paleontology to the End of the Nineteenth Century.* London: W. Scott.

Index

Index